P9-EMH-967

ALSO BY OLIVER SACKS

Migraine

Awakenings

A Leg to Stand On

The Man Who Mistook His Wife for a Hat

Seeing Voices:
A Journey into the World of the Deaf

An Anthropologist
on Mars

An
Anthropologist
on Mars

SEVEN PARADOXICAL TALES

Oliver Sacks

ALFRED A. KNOPF
NEW YORK
1995

THIS IS A BORZOI BOOK
PUBLISHED BY ALFRED A. KNOPF, INC.

Copyright © 1995 by Oliver Sacks
All rights reserved under International and Pan-American
Copyright Conventions. Published in the United States
by Alfred A. Knopf, Inc., New York, and simultaneously
in Canada by Random House of Canada Limited, Toronto.
Distributed by Random House, Inc., New York.

Earlier versions of the essays in this work were originally
published in the following:
The New York Review of Books: "The Case of the
Colorblind Painter" and "The Last Hippie"
The New Yorker: "An Anthropologist on Mars," "The
Landscape of His Dreams," "Prodigies," "To See and Not
See," and "A Surgeon's Life"

Owing to limitations of space, permission to reprint previously published material may be found following the
index.

Library of Congress Cataloging-in-Publication Data

Sacks, Oliver W.
    An anthropologist on Mars : 7 paradoxical tales / Oliver
Sacks.—1st ed.
    p.   cm.
    Includes bibliographical references (p.   ).
    ISBN 0-679-43785-1
    1. Neurology—Anecdotes.   I. Title.
RC351.S1948   1995
616.8   DC20
                                            94-26733
                                                CIP

Manufactured in the United States of America
First Edition

To the seven
whose stories are related here

The universe is not only queerer than we imagine, but queerer than we can imagine.

J. B. S. Haldane

Ask not what disease the person has, but rather what person the disease has.

(attributed to) William Osler

Contents

Acknowledgments

First, I am deeply grateful to my subjects: "Jonathan I.," "Greg F.," "Carl Bennett," "Virgil," Franco Magnani, Stephen Wiltshire, and Temple Grandin. To them, their families, their friends, their physicians and therapists, I owe an infinite debt.

Two very special colleagues have been Bob Wasserman (who was my co-author on the original version of "The Case of the Colorblind Painter") and Ralph Siegel (who has been a collaborator in other books)—we formed a sort of team in the cases of Jonathan I. and Virgil.

I owe to many friends and colleagues (more than I can enumerate!) information, help, and stimulating discussion. With some there has been a close, continual colloquy over the years, as with Jerry Bruner and Gerald Edelman; with others only occasional meetings and letters; but all have excited and inspired me in different ways. These include: Ursula Bellugi, Peter Brook, Jerome Bruner, Elizabeth Chase, Patricia and Paul Churchland, Joanne Cohen, Pietro Corsi, Francis Crick, Antonio and Hanna Damasio, Merlin Donald, Freeman Dyson, Gerald Edelman, Carol Feldman, Shane Fistell, Allen Furbeck, Frances Futterman, Elkhonon Goldberg, Stephen Jay Gould, Richard Gregory, Kevin Halligan, Lowell Handler, Mickey Hart, Jay Itzkowitz, Helen Jones, Eric Korn, Deborah Lai, Skip and Doris Lane, Sue Levi-Pearl, John MacGregor, John Marshall, Juan Martinez, Jonathan and Rachel Miller, Arnold Modell, Jonathan Mueller, Jock Murray, Knut Nordby,

Acknowledgments

Michael Pearce, V. S. Ramachandran, Isabelle Rapin, Chris Rawlence, Bob Rodman, Israel Rosenfield, Carmel Ross, Yolanda Rueda, David Sacks, Marcus Sacks, Michael Sacks, Dan Schachter, Murray Schane, Herb Schaumburg, Susan Schwartzenberg, Robert Scott, Richard Shaw, Leonard Shengold, Larry Squire, John Steele, Richard Stern, Deborah Tannen, Esther Thelen, Connie Tomaino, Russell Warren, Ed Weinberger, Ren and Joasia Weschler, Andrew Wilkes, Jerry Young, Semir Zeki.

Many people have shared their knowledge and expertise in the field of autism with me, including, first and foremost, my good friend and colleague Isabelle Rapin, Doris Allen, Howard Bloom, Marlene Breitenbach, Uta Frith, Denise Fruchter, Beate Hermelin, Patricia Krantz, Lynn McClannahan, Clara and David Park, Jessy Park, Sally Ramsey, Ginger Richardson, Bernard Rimland, Ed and Riva Ritvo, Mira Rothenberg, and Rosalie Winard. In relation to Stephen Wiltshire, I must thank Lorraine Cole, Chris Marris, and above all, Margaret and Andrew Hewson.

I am grateful to innumerable correspondents (including the now-unknown correspondent who sent me a copy of the 1862 Fayetteville *Observer*), some of whom are quoted in these pages. Many of these explorations, indeed, started with unexpected letters or phone calls, beginning with Mr. I.'s letter to me in March 1986.

There are places, no less than people, that have contributed to this book, by providing shelter, calm, stimulation. Foremost among them has been the New York Botanical Garden (and especially the now-dismantled fern collection), my favorite place for walking and thinking; the Lake Jefferson Hotel and its lake; Blue Mountain Center (and Harriet Barlow); the New York Institute for the Humanities, where some of the testing of Mr. I. was done; the library at the Albert Einstein College of Medicine, which has helped me track down many sources; and lakes, rivers, and swimming pools everywhere—for I do most of my thinking in the water.

The Guggenheim Foundation very generously supported my work on "A Surgeon's Life" with a 1989 grant for research on the neuroanthropology of Tourette's syndrome.

Earlier versions of "The Case of the Colorblind Painter" and "The Last Hippie" were published in *The New York Review of Books*, and of the other case histories in *The New Yorker*. I have been privileged to have worked with Robert Silvers at the *NYRB*, and John Bennet at *The New Yorker*, and the staff of both publications. Many others have contributed to the editing and publication of this book, including Dan Frank and Claudine O'Hearn at Knopf, Jacqui Graham at Picador, Jim Silberman, Heather Schroder, Susan Jensen, and Suzanne Gluck. Finally, someone who has known all the subjects in this book, and has helped to give it impetus and shape, has been my assistant, editor, collaborator, and friend, Kate Edgar.

But to return to where I started—for all clinical studies, however widely they adventure, or deeply they investigate, must return to the concrete subject, the individuals who inspired them, and whom they are about. So to the seven people who have trusted me, shared their lives with me, given so deeply of their own experience—and who, over the years, have become my friends—I dedicate this book.

Preface

I am writing this with my left hand, although I am strongly right-handed. I had surgery to my right shoulder a month ago and am not permitted, not capable of, use of the right arm at this time. I write slowly, awkwardly—but more easily, more naturally, with each passing day. I am adapting, learning, all the while—not merely this left-handed writing, but a dozen other left-handed skills as well: I have also become very adept, prehensile, with my toes, to compensate for having one arm in a sling. I was quite off balance for a few days when the arm was first immobilized, but now I walk differently, I have discovered a new balance. I am developing different patterns, different habits . . . a different identity, one might say, at least in this particular sphere. There must be changes going on with some of the programs and circuits in my brain—altering synaptic weights and connectivities and signals (though our methods of brain imaging are still too crude to show these).

Though some of my adaptations are deliberate, planned, and some are learned through trial and error (in the first week I injured every finger of my left hand), most have occurred by themselves, unconsciously, by reprogrammings and adaptations of which I know nothing (any more than I know, or can know, how I normally walk). Next month, if all goes well, I can start to readapt again, to regain a full (and "natural") use of the right arm, to reincorporate it back into my body image, myself, to become a dexterous, dextral human being once again.

But recovery, in such circumstances, is by no means automatic, a simple process like tissue healing—it will involve a whole nexus of muscular and postural adjustments, a whole sequence of new procedures (and their synthesis), learning, finding, a new path to recovery. My surgeon, an understanding man who has had the same operation himself, said, "There are *general* guidelines, restrictions, recommendations. But all the particulars you will have to find out for yourself." Jay, my physiotherapist, expressed himself similarly: "Adaptation follows a different path in each person. The nervous system creates its own paths. You're the neurologist—you must see this all the time."

Nature's imagination, as Freeman Dyson likes to say, is richer than ours, and he speaks, marvellingly, of this richness in the physical and biological worlds, the endless diversity of physical forms and forms of life. For me, as a physician, nature's richness is to be studied in the phenomena of health and disease, in the endless forms of individual adaptation by which human organisms, people, adapt and reconstruct themselves, faced with the challenges and vicissitudes of life.

Defects, disorders, diseases, in this sense, can play a paradoxical role, by bringing out latent powers, developments, evolutions, forms of life, that might never be seen, or even be imaginable, in their absence. It is the paradox of disease, in this sense, its "creative" potential, that forms the central theme of this book.

Thus while one may be horrified by the ravages of developmental disorder or disease, one may sometimes see them as creative too—for if they destroy particular paths, particular ways of doing things, they may force the nervous system into making other paths and ways, force on it an unexpected growth and evolution. This other side of development or disease is something I see, potentially, in almost every patient; and it is this, here, which I am especially concerned to describe.

Similar considerations were brought up by A. R. Luria, who, more than any other neurologist in his lifetime, studied the long-term survival of patients who had cerebral tumors or had suffered brain injuries or strokes—and the ways, the adaptations, they used to survive. He also studied deaf and blind children as a very young man (with his mentor L. S. Vygotsky). Vygotsky stressed the intactness rather than the deficits of such children:

> A handicapped child represents a qualitatively different, unique type of development. . . . If a blind or deaf child achieves the same level of development as a normal child, then the child with a defect achieves this *in another way, by another course, by other means*; and, for the pedagogue, it is particularly important to know the uniqueness of the course along which he must lead the child. This uniqueness transforms the minus of the handicap into the plus of compensation.

That such radical adaptations could occur demanded, Luria thought, a new view of the brain, a sense of it not as programmed and static, but rather as dynamic and active, a supremely efficient adaptive system geared for evolution and change, ceaselessly adapting to the needs of the organism—its need, above all, to construct a coherent self and world, whatever defects or disorders of brain function befell it. That the brain is minutely differentiated is clear: there are hundreds of tiny areas crucial for every aspect of perception and behavior (from the perception of color and of motion to, perhaps, the intellectual orientation of the individual). The miracle is how they all cooperate, are integrated together, in the creation of a self.[1]

---

[1] This, indeed, is *the* problem, the ultimate question, in neuroscience—and it cannot be answered, even in principle, without a global theory of brain function, one capable of showing the interactions of every level, from the micropatterns of individual neuronal responses to the grand macropatterns of an actual lived life. Such a theory, a neural theory of personal identity, has been proposed in the last few years by Gerald M. Edelman, in his theory of neuronal group selection, or "neural Darwinism."

This sense of the brain's remarkable plasticity, its capacity for the most striking adaptations, not least in the special (and often desperate) circumstances of neural or sensory mishap, has come to dominate my own perception of my patients and their lives. So much so, indeed, that I am sometimes moved to wonder whether it may not be necessary to redefine the very concepts of "health" and "disease," to see these in terms of the ability of the organism to create a new organization and order, one that fits its special, altered disposition and needs, rather than in the terms of a rigidly defined "norm."

Sickness implies a contraction of life, but such contractions do not have to occur. Nearly all of my patients, so it seems to me, whatever their problems, reach out to life—and not only despite their conditions, but often because of them, and even with their aid.

Here then are seven narratives of nature—and the human spirit—as these have collided in unexpected ways. The people in this book have been visited by neurological conditions as diverse as Tourette's syndrome, autism, amnesia, and total colorblindness. They exemplify these conditions, they are "cases" in the traditional medical sense—but equally they are unique individuals, each of whom inhabits (and in a sense has created) a world of his own.

These are tales of survival, survival under altered, sometimes radically altered, conditions—survival made possible by the wonderful (but sometimes dangerous) powers of reconstruction and adaptation we have. In earlier books I wrote of the "preservation" of self, and (more rarely) of the "loss" of self, in neurological disorders. I have to come to think these terms too simple—and that there is neither loss nor preservation of identity in such situations, but, rather, its adaptation, even its transmutation, given a radically altered brain and "reality."

The study of disease, for the physician, demands the study of identity, the inner worlds that patients, under the spur of illness, create. But the realities of patients, the ways in which they and their brains construct their own worlds, cannot be comprehended wholly from the observation of behavior, from the outside. In addition to the objective approach of the scientist, the naturalist, we must employ an intersubjective approach too, leaping, as Foucault writes, "into the interior of morbid consciousness, [trying] to see the pathological world with the eyes of the patient himself." No one has written better of the nature and necessity of such intuition or empathy than G. K. Chesterton, through the mouth of his spiritual detective, Father Brown. Thus when Father Brown is asked for his method, his secret, he replies:

> Science is a grand thing when you can get it; in its real sense one of the grandest words in the world. But what do these men mean, nine times out of ten, when they use it nowadays? When they say detection is a science? When they say criminology is a science? They mean getting *outside* a man and studying him as if he were a gigantic insect; in what they would call a dry impartial light; in what I should call a dead and dehumanized light. They mean getting a long way off him, as if he were a distant prehistoric monster; staring at the shape of his "criminal skull" as if it were a sort of eerie growth, like the horn on a rhinoceros's nose. When the scientist talks about a type, he never means himself, but always his neighbour; probably his poorer neighbour. I don't deny the dry light may sometimes do good; though in one sense it's the very reverse of science. So far from being knowledge, it's actually suppression of what we know. It's treating a friend as a stranger, and pretending that something familiar is really remote and mysterious. It's like saying that a man has a proboscis between the eyes, or that he falls down in a fit of insensibility once every twenty-four hours. Well, what you call "the secret" is exactly the opposite. I don't try to get outside the man. I try to get inside.

The exploration of deeply altered selves and worlds is not one that can be fully made in a consulting room or office. The

Preface

French neurologist François Lhermitte is especially sensitive to this, and instead of just observing his patients in the clinic, he makes a point of visiting them at home, taking them to restaurants or theaters, or for rides in his car, sharing their lives as much as possible. (It is similar, or was similar, with physicians in general practice. Thus when my father was reluctantly considering retirement at ninety, we said, "At least drop the house calls." But he answered, "No, I'll keep the house calls—I'll drop everything else instead.")

With this in mind, I have taken off my white coat, deserted, by and large, the hospitals where I have spent the last twenty-five years, to explore my subjects' lives as they live in the real world, feeling in part like a naturalist, examining rare forms of life; in part like an anthropologist, a neuroanthropologist, in the field—but most of all like a physician, called here and there to make house calls, house calls at the far borders of human experience.

These then are tales of metamorphosis, brought about by neurological chance, but metamorphosis into alternative states of being, other forms of life, no less human for being so different.

New York   O.W.S.
June 1994

An Anthropologist
on Mars

The Case of the Colorblind Painter

Early in March 1986 I received the following letter:

I am a rather successful artist just past 65 years of age. On January 2nd of this year I was driving my car and was hit by a small truck on the passenger side of my vehicle. When visiting the emergency room of a local hospital, I was told I had a concussion. While taking an eye examination, it was discovered that I was unable to distinguish letters or colors. The letters appeared to be Greek letters. My vision was such that everything appeared to me as viewing a black and white television screen. Within days, I could distinguish letters and my vision became that of an eagle—I can see a worm wriggling a block away. The sharpness of focus is incredible. BUT—I AM ABSOLUTELY COLOR BLIND. I have visited ophthalmologists who know nothing about this color-blind business. I have visited neurologists, to no avail. Under hypnosis I still can't distinguish colors. I have been involved in all kinds of tests. You name it. My brown dog is dark grey. Tomato juice is black. Color TV is a hodge-podge. . . .

Had I ever encountered such a problem before, the writer continued; could I explain what was happening to him—and could I help?

This seemed an extraordinary letter. Colorblindness, as ordinarily understood, is something one is born with—a difficulty distinguishing red and green, or other colors, or (extremely rarely) an inability to see any colors at all, due to defects in the color-responding cells, the cones, of the retina. But clearly this was not the case with my correspondent, Jonathan I. He had seen normally all his life, had been born with a full complement of cones in his retinas. He had *become* colorblind, after sixty-five years of seeing colors normally— *totally* colorblind, as if "viewing a black and white television screen." The suddenness of the event was incompatible with any of the slow deteriorations that can befall the retinal cone cells and suggested instead a mishap at a much higher level, in those parts of the brain specialized for the perception of color.

Total colorblindness caused by brain damage, so-called cerebral achromatopsia, though described more than three centuries ago, remains a rare and important condition. It has intrigued neurologists because, like all neural dissolutions and destructions, it can reveal to us the mechanisms of neural construction—specifically, here, how the brain "sees" (or makes) color. Doubly intriguing is its occurrence in an artist, a painter for whom color has been of primary importance, and who can directly paint as well as describe what has befallen him, and thus convey the full strangeness, distress, and reality of the condition.

Color is not a trivial subject but one that has compelled, for hundreds of years, a passionate curiosity in the greatest artists, philosophers, and natural scientists. The young Spinoza wrote his first treatise on the rainbow; the young Newton's most joyous discovery was the composition of white light; Goethe's great color work, like Newton's, started with a prism; Schopenhauer, Young, Helmholtz, and Maxwell, in the last century, were all tantalized by the problem of color; and Wittgenstein's last work was his *Remarks on Colour*. And yet most of us, most of the time, overlook its great mystery. Through such a case as Mr. I.'s we can trace not only the un-

derlying cerebral mechanisms or physiology but the phenom-
enology of color and the depth of its resonance and meaning
for the individual.

On getting Mr. I.'s letter, I contacted my good friend and
colleague Robert Wasserman, an ophthalmologist, feeling that
together we needed to explore Mr. I.'s complex situation and,
if we could, help him. We first saw him in April 1986. He was
a tall, gaunt man, with a sharp, intelligent face. Although obviously depressed by his condition, he soon warmed to us and
began talking with animation and humor. He constantly
smoked as he talked; his fingers, restless, were stained with
nicotine. He described a very active and productive life as
an artist, from his early days with Georgia O'Keeffe in New
Mexico, to painting backdrops in Hollywood during the
1940s, to working as an Abstract Expressionist in New York
during the 1950s and later as an art director and a commer-
cial artist.

We learned that his accident had been accompanied by a
transient amnesia. He had been able, evidently, to give a clear
account of himself and his accident to the police at the time
it happened, late on the afternoon of January 2, but then, because of a steadily mounting headache, he went home. He
complained to his wife of having a headache and feeling confused, but made no mention of the accident. He then fell into
a long, almost stuporous sleep. It was only the next morning,
when his wife saw the side of the car stove in, that she asked
him what had happened. When she got no clear answer ("I
don't know. Maybe somebody backed into it") she knew that
something serious must have happened.

Mr. I. then drove off to his studio and found on his desk a
carbon copy of the police accident report. He had had an accident, but somehow, bizarrely, had lost his memory of it. Perhaps the report would jolt his memory. But lifting it up, he
could make nothing of it. He saw print of different sizes and
types, all clearly in focus, but it looked like "Greek" or "He-

brew" to him.[1] A magnifying glass did not help; it simply became *large* "Greek" or "Hebrew." (This alexia, or inability to read, lasted for five days, but then disappeared.)

Feeling now that he must have suffered a stroke or some sort of brain damage from the accident, Jonathan I. phoned his doctor, who arranged for him to be tested at a local hospital. Although, as his original letter indicates, difficulties in distinguishing colors were detected at this time, in addition to his inability to read, he had no subjective sense of the alteration of colors until the next day.

That day he decided to go to work again. It seemed to him as if he were driving in a fog, even though he knew it to be a bright and sunny morning. Everything seemed misty, bleached, greyish, indistinct. He was flagged down by the police close to his studio: he had gone through two red lights, they said. Did he realize this? No, he said, he was not aware of having passed through any red lights. They asked him to get out of the car. Finding him sober, but apparently bewildered and ill, they gave him a ticket and suggested he seek medical advice.

Mr. I. arrived at his studio with relief, expecting that the horrible mist would be gone, that everything would be clear again. But as soon as he entered, he found his entire studio, which was hung with brilliantly colored paintings, now utterly grey and void of color. His canvases, the abstract color paintings he was known for, were now greyish or black and white. His paintings—once rich with associations, feelings, meanings—now looked unfamiliar and meaningless to him. At this point the magnitude of his loss overwhelmed him. He had spent his life as a painter; now even his art was without meaning, and he could no longer imagine how to go on.

The weeks that followed were very difficult. "You might

---

[1] I asked Mr. I. later if he knew Greek or Hebrew; he said no, there was just the sense of an unintelligible foreign language; perhaps, he added, "cuneiform" would be more accurate. He saw forms, he knew they had to have meaning, but could not imagine what this meaning might be.

think," Mr. I. said, "loss of color vision, what's the big deal? Some of my friends said this, my wife sometimes thought this, but to me, at least, it was awful, disgusting." He *knew* the colors of everything, with an extraordinary exactness (he could give not only the names but the numbers of colors as these were listed in a Pantone chart of hues he had used for many years). He could identify the green of van Gogh's billiard table in this way unhesitatingly. He *knew* all the colors in his favorite paintings, but could no longer see them, either when he looked or in his mind's eye. Perhaps he knew them, now, only by verbal memory.

It was not just that colors were missing, but that what he did see had a distasteful, "dirty" look, the whites glaring, yet discolored and off-white, the blacks cavernous—everything wrong, unnatural, stained, and impure.[2]

Mr. I. could hardly bear the changed appearances of people ("like animated grey statues") any more than he could bear his own appearance in the mirror: he shunned social intercourse and found sexual intercourse impossible. He saw people's flesh, his wife's flesh, his own flesh, as an abhorrent grey; "flesh-colored" now appeared "rat-colored" to him. This was so even when he closed his eyes, for his vivid visual imagery was preserved but was now without color as well.

The "wrongness" of everything was disturbing, even disgusting, and applied to every circumstance of daily life. He found foods disgusting due to their greyish, dead appearance and had to close his eyes to eat. But this did not help very much, for the mental image of a tomato was as black as its appearance. Thus, unable to rectify even the inner image, the idea, of various foods, he turned increasingly to black and white foods—to black olives and white rice, black coffee and yogurt. These at least appeared relatively normal, whereas most foods, normally colored, now appeared horribly abnor-

---

[2] Similarly, a patient of Dr. Antonio Damasio, with achromatopsia from a tumor, thought everything and everyone looked "dirty," even finding new-fallen snow unpleasant and dirty.

mal. His own brown dog looked so strange to him now that he even considered getting a Dalmatian.

He encountered difficulties and distresses of every sort, from the confusion of red and green traffic lights (which he could now distinguish only by position) to an inability to choose his clothes. (His wife had to pick them out, and this dependency he found hard to bear; later, he had everything classified in his drawers and closet—grey socks here, yellow there, ties labeled, jackets and suits categorized, to prevent otherwise glaring incongruities and confusions.) Fixed and ritualistic practices and positions had to be adopted at the table; otherwise he might mistake the mustard for the mayonnaise, or, if he could bring himself to use the blackish stuff, ketchup for jam.[3]

---

[3] In 1688, in *Some Uncommon Observations about Vitiated Sight*, Robert Boyle described a young woman in her early twenties whose eyesight had been normal until she was eighteen, when she developed a fever, was "tormented with blisters," and, with this, "deprived of her sight." When she was presented with something red, "she look'd attentively upon it, but told me, that to her, it did not seem Red, but of another Colour, which one would guess by her description to be a Dark or Dirty one." When "tufts of Silk that were finely Color'd" were given to her, she could only say that "they seem'd to be a Light-colour, but could not tell which." When asked whether the meadows "did not appear to her Cloathed in Green," she said they did not, but seemed to be "of an odd Darkish colour," adding that when she wished to gather violets, "she was not able to distinguish them by the Colour from the surrounding Grass, but only by the Shape, or by feeling them." Boyle further observed a change in her habits, that she liked now to walk abroad in the evenings, and this "she much delighteth to do."

A number of accounts were published in the nineteenth century—many collected in Mary Collins's *Colour-Blindness*—one of the most vivid (besides that of an achromatopic house painter) being that of a physician who, thrown from his horse, suffered a head injury and concussion. "On recovering sufficiently to notice objects around him," George Wilson recorded in 1853,

he found that his perception of colours, which was formerly normal and acute, had become both weakened and perverted. . . . All coloured objects . . . now seem strange to him. . . . Whilst formerly a student in Edinburgh he was known as an excellent anatomist; now he cannot distinguish an artery from a vein by its tint. . . . Flowers have lost more than half their beauty for him, and he recalls the shock which he received on first entering his garden after his recovery, at finding that a favourite damask rose had become in all its parts,

As the months went by, he particularly missed the brilliant colors of spring—he had always loved flowers, but now he could only distinguish them by shape or smell. The blue jays were brilliant no longer; their blue, curiously, was now seen as pale grey. He could no longer see the clouds in the sky, their whiteness, or off-whiteness as he saw them, being scarcely distinguishable from the azure, which seemed bleached to a pale grey. Red and green peppers were also indistinguishable, but this was because both appeared black. Yellows and blues, to him, were almost white.[4]

Mr. I. also seemed to experience an excessive tonal contrast, with loss of delicate tonal gradations, especially in direct sunlight or harsh artificial light; he made a comparison here with the effects of sodium lighting, which at once removes color and tonal delicacy, and with certain black-and-white films—"like Tri-X pushed for speed"—which produce a harsh, contrasty effect. Sometimes objects stood out with inordinate contrast and sharpness, like silhouettes. But if the contrast was normal, or low, they might disappear from sight altogether.

Thus, though his brown dog would stand out sharply in silhouette against a light road, it might get lost to sight when it moved into soft, dappled undergrowth. People's figures might be visible and recognizable half a mile off (as he himself said

---

petals, leaves, and stem, of one uniform dull colour; and that variegated flowers had lost their characteristic tints.

[4] One sees interesting similarities, but also differences, from the vision of those with congenital achromatopsia. Thus Knut Nordby, a congenitally colorblind vision researcher, writes:

I only see the world in shades that colour-normals describe as black, white and grey. My subjective spectral sensitivity is not unlike that of orthochromatic black and white film. I experience the colour called red as a very dark grey, nearly black, even in very bright light. On a grey-scale the blue and green colours I see as mid-greys, somewhat darker greys if they are saturated, somewhat lighter greys when unsaturated. Yellow typically appears to me as a rather light grey, but is usually not confused with white. Brown usually appears as a dark grey and so does a very saturated orange.

in his original letter, and many times later, his vision had become much sharper, "that of an eagle"), but faces would often be unidentifiable until they were close. This seemed a matter of lost color and tonal contrast, rather than a defect in recognition, an agnosia. A major problem occurred when he drove, in that he tended to misinterpret shadows as cracks or ruts in the road and would brake or swerve suddenly to avoid these.

He found color television especially hard to bear: its images always unpleasant, sometimes unintelligible. Black-and-white television, he thought, was much easier to deal with; he felt his perception of black-and-white images to be relatively normal, whereas something bizarre and intolerable occurred whenever he looked at colored images. (When we asked why he did not simply turn off the color, he said he thought that the tonal values of "decolored" color TV seemed different, less "normal," than those of a "pure" black-and-white set.) But, as he now explained, in distinction to his first letter, his world was not really like black-and-white television or film—it would have been much easier to live with had it been so. (He sometimes wished he could wear miniature TV glasses.)

His despair of conveying what his world looked like, and the uselessness of the usual black-and-white analogies, finally drove him, some weeks later, to create an entire grey room, a grey universe, in his studio, in which tables, chairs, and an elaborate dinner ready for serving were all painted in a range of greys. The effect of this, in three dimensions and in a different tonal scale from the "black and white" we are all accustomed to, was indeed macabre, and wholly unlike that of a black-and-white photograph. As Mr. I. pointed out, we accept black-and-white photographs or films because they are *representations* of the world—images that we can look at, or away from, when we want. But black and white for him was a *reality*, all around him, 360 degrees, solid and three-dimensional, twenty-four hours a day. The only way he could express it, he felt, was to make a completely grey room for others to experience—but of course, he pointed out, the observer himself would have to be painted grey, so he would be

Two paintings done by
Mr. I. shortly before his
accident.

A painting of flowers done four weeks after Mr. I. 's accident. The underlying outlines are clear, but camouflaged by a random application of color.

Mr. I. painted pieces of grey fruit to show us the "leaden" universe into which he had fallen.

A test painting from Mary Collins's *Colour-Blindness* (left), as reproduced by someone with red-green colorblindness, and by Mr. I. (right).

The sunset scene of which Mr. I. could see virtually nothing—an effect simulated by a black-and-white photocopy of it.

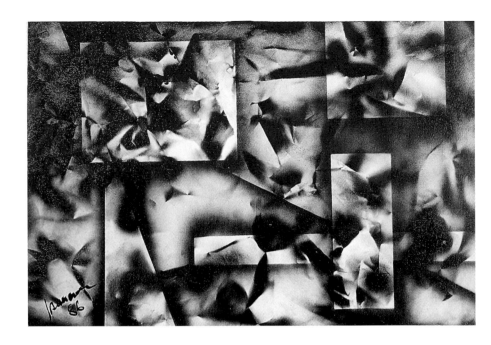

A black-and-white painting done about two months after Mr. I.'s accident, and a painting done two years later—Mr. I. at this time was experimenting with adding single colors, even though he could not see them.

part of the world, not just observing it. More than this: the observer would have to lose, as he himself had, the neural knowledge of color. It was, he said, like living in a world "molded in lead."

Subsequently, he said neither "grey" nor "leaden" could begin to convey what his world was actually like. It was not "grey" that he experienced, he said, but perceptual qualities for which ordinary experience, ordinary language, had no equivalent.

Mr. I. could no longer bear to go to museums and galleries or to see colored reproductions of his favorite pictures. This was not just because they were bereft of color, but because they looked intolerably *wrong*, with washed-out or "unnatural" shades of grey (photographs in black and white, on the other hand, were much more tolerable). This was especially distressing when he knew the artists, and the perceptual debasement of their work interfered with his sense of their identity—this, indeed, was what he now felt was happening with himself.

He was depressed once by a rainbow, which he saw only as a colorless semicircle in the sky. And he even felt his occasional migraines as "dull"—previously they had involved brilliantly colored geometric hallucinations, but now even these were devoid of color. He sometimes tried to evoke color by pressing the globes of his eyes, but the flashes and patterns elicited were equally lacking in color. He had often dreamed in vivid color, especially when he dreamed of landscapes and painting; now his dreams were washed-out and pale, or violent and contrasty, lacking both color and delicate tonal gradations.

Music, curiously, was impaired for him too, because he had previously had an extremely intense synesthesia, so that different tones had immediately been translated into color, and he experienced all music simultaneously as a rich tumult of inner colors. With the loss of his ability to generate colors, he lost this ability as well—his internal "color-organ" was out of action, and now he heard music with no visual accompaniment; this, for him, was music with its essential chromatic counterpart missing, music now radically impoverished.[5]

A certain mild pleasure came from looking at drawings; he had been a fine draftsman in his earlier years. Could he not go back to drawing again? This thought was slow to occur to him, and it only took hold after being suggested repeatedly by others. His own first impulse was to paint in color. He insisted that he still "knew" what colors to use, even though he could no longer see them. He decided, as a first exercise, to paint flowers, taking from his palette what tints seemed "tonally right." But the pictures were unintelligible, a confusing welter of colors to normal eyes. It was only when one of his artist friends took black-and-white Polaroids of the paintings that they made sense. The contours were accurate, but the colors were all wrong. "No one will get your paintings," one of his friends said, "unless they are as colorblind as you."

"Stop pushing it," said another. "You *can't* use color now." Mr. I. reluctantly allowed all his colored paints to be put away. It's only temporary, he thought. I'll be back to color soon.

These first weeks were a time of agitation, even desperation; he was constantly hoping that he would wake up one fine morning and find the world of color miraculously restored. This was a constant motif in his dreams at the time, but the wish was never fulfilled, even in his dreams. He would dream that he was *about* to see in color, but then he would wake and find that nothing had changed. He constantly feared that whatever had happened would happen again, this time depriving him of all his sight completely. He thought he had probably had a stroke, caused by (or perhaps causing) his accident in the car, and feared that there could be another stroke

---

[5] Only one sense could give him any real pleasure at this time, and this was the sense of smell. Mr. I. had always had a most acute, erotically charged sense of smell—indeed, he ran a small perfume business on the side, compounding his own scents. As the pleasures of seeing were lost, the pleasures of smell were heightened (or so it seemed to him), in the first grim weeks after his accident.

at any moment. In addition to this medical fear, there was a deeper bewilderment and fear that he found almost impossible to articulate, and it was this that had come to a head in his month of attempted color painting, his month of insisting that he still "knew" color. It had gradually come upon him, during this time, that it was not merely color perception and color imagery that he lacked, but something deeper and difficult to define. He knew all about color, externally, intellectually, but he had lost the remembrance, the inner knowledge, of it that had been part of his very being. He had had a lifetime of experience in color, but now this was only a historical fact, not something he could access and feel directly. It was as if his past, his chromatic past, had been taken away, as if the brain's knowledge of color had been totally excised, leaving no trace, no inner evidence, of its existence behind.[6]

By the beginning of February, some of his agitation was calming down; he had started to accept, not merely intellectually, but at a deeper level, too, that he was indeed totally colorblind and might possibly remain so. His initial sense of helplessness started to give way to a sense of resolution—he would paint in black and white, if he could not paint in color; he would try to live in a black-and-white world as fully as he could. This resolution was strengthened by a singular experi-

---

[6] The question of "knowing" color is very complex and has paradoxical aspects that are difficult to dissect. Certainly Mr. I. was intensely aware of a profound loss with the change in his vision, so clearly some sort of comparison with past experience was possible for him. Such a comparison is not possible if there is a complete destruction of the primary visual cortex on both sides, say from a stroke, as in Anton's syndrome. Patients with this syndrome become totally blind, but make no complaint or report of their blindness. They do not know they are blind; the whole structure of consciousness is completely reorganized—instantly so—at the moment they are stricken.

Similarly, patients with massive strokes in the right parietal cortex may lose not only the sensation and use but the very knowledge of their left sides, of everything to the left, and indeed of the very concept of leftness. But they are "anosognosic"—they have no knowledge of their loss; *we* may say their world is bisected, but, for them, it is whole and complete.

ence, about five weeks after his accident, as he was driving to the studio one morning. He saw the sunrise over the highway, the blazing reds all turned into black: "The sun rose like a bomb, like some enormous nuclear explosion," he said later. "Had anyone ever seen a sunrise in this way before?"

Inspired by the sunrise, he started painting again—he started, indeed, with a black-and-white painting that he called *Nuclear Sunrise*, and then went on to the abstracts he favored, but now painting in black and white only. The fear of blindness continued to haunt him but, creatively transmuted, shaped the first "real" paintings he did after his color experiments. Black-and-white paintings he now found he could do, and do very well. He found his only solace working in the studio, and he worked fifteen, even eighteen, hours a day. This meant for him a kind of artistic survival: "I felt if I couldn't go on painting," he said later, "I wouldn't want to go on at all."

His first black-and-white paintings, done in February and March, gave a feeling of violent forces—rage, fear, despair, excitement—but these were held in control, attesting to the powers of artistry that could disclose, and yet contain, such intensity of feeling. In these two months he produced dozens of paintings, marked by a singular style, a character he had never shown before. In many of these paintings, there was an extraordinary shattered, kaleidoscopic surface, with abstract shapes suggestive of faces—averted, shadowed, sorrowing, raging—and dismembered body parts, faceted and held in frames and boxes. They had, compared with his previous work, a labyrinthine complexity, and an obsessed, haunted quality—they seemed to exhibit, in symbolic form, the predicament he was in.

Starting in May—it was fascinating to watch—he moved from these powerful but rather terrifying and alien paintings toward themes, living themes, he had not touched in thirty years, back to representational paintings of dancers and racehorses. These paintings, even though still in black and white, were full of movement, vitality, and sensuousness; and they

went with a change in his personal life—a lessening of his
withdrawal and the beginnings of a renewed social and sexual
life, a lessening of his fears and depression, and a turning back
to life.

At this time, too, he turned to sculpture, which he had
never done before. He seemed to be turning to all the visual
modes that still remained to him—form, contour, movement,
depth—and exploring them with heightened intensity. He also
started painting portraits, although he found that he could not
work from life, but only from black-and-white photographs,
fortified by his knowledge of and feeling for each subject. Life
was tolerable only in the studio, for here he could reconceive
the world in powerful, stark forms. But outside, in real life, he
found the world alien, empty, dead, and grey.

This was the story Bob Wasserman and I got from Mr. I.—a
story of an abrupt and total breakdown of color vision, and his
attempts to live in a black-and-white world. I had never been
given such a history before, I had never met anyone with total
colorblindness before, and I had no idea what had happened to
him—nor whether his condition could be reversed or improved.

The first thing was to define his impairments more precisely with various tests, some quite informal, making use of
everyday objects or pictures, whatever came to hand. For instance, we first asked Mr. I. about a shelf of notebooks—blue,
red, and black—by my desk. He instantly picked out the blue
ones (a bright medium blue to normal eyes)—"they're pale."
The red and the black were indistinguishable—both, for him,
were "dead black."

We then gave him a large mass of yarns, containing thirty-
three separate colors, and asked him to sort these: he said he
could not sort them by color, but only by grey-scale tonal values. He then, rapidly and easily, separated the yarns into four
strange, chromatically random piles, which he characterized as
0–25 percent, 25–50 percent, 50–75 percent, and 75–100 percent
on a grey-tone scale (though nothing looked to him purely
white, and even white yarn looked slightly "dingy" or "dirty").

We ourselves could not confirm the accuracy of this, because our color vision interfered with our ability to visualize a grey scale, just as normally sighted viewers had been unable to perceive the tonal sense of his confusingly polychromatic flower paintings. But a black-and-white photograph and a black-and-white video camera confirmed that Mr. I. had indeed accurately divided the colored yarns in a grey scale that basically coincided with their own mechanical reading. There was, perhaps, a certain crudeness in his categories, but this went with the sense of sharp contrast, the paucity of tonal gradations, that he had complained of. Indeed, when shown an artist's grey scale of perhaps a dozen gradations from black to white, Mr. I. could distinguish only three or four categories of tone.[7]

We also showed him the classic Ishihara color-dot plates, in which configurations of numerals in subtly differentiated colors may stand out clearly for the normally sighted, but not for those with various types of colorblindness. Mr. I. was unable to see any of these figures (although he was able to see certain plates that are visible to the colorblind but not to normally sighted people, and thus designed to catch pretended or hysterical colorblindness).[8]

---

[7] One anomaly showed itself in the yarn-sorting test; he ranked bright saturated blues as "pale" (as he had complained that the blue sky seemed almost white). But was this an anomaly? Could we be sure that the blue wool was not, under its blueness, rather washed-out or pale? We had to have hues that were otherwise identical—identical in brightness, saturation, reflectivity, so we obtained a set of carefully produced color buttons known as the Farnsworth-Munsell test and gave this to Mr. I. He was unable to put the buttons in any order, but he did separate out the blue ones as "paler" than the rest.

[8] Further testing with the Nagel anomaloscope and the Sloan achromatopsia cards confirmed Mr. I.'s total colorblindness. With Dr. Ralph Siegel, we did tests of depth and motion perception (using Julesz random-dot stereograms and moving random-dot fields)—these were normal, as were tests of his ability to generate structure and depth from motion. There was, however, one interesting anomaly: Mr. I. was unable to "get" red and green stereograms (bicolor anaglyphs), presumably because color vision is needed to segregate the two images. We also obtained electroretinograms, and these were quite normal, indicating that all three cone mechanisms in the retina were intact, and that the colorblindness was indeed of cerebral origin.

We happened to have a postcard that could have been designed for testing achromatopes—a postcard of a coastal scene, with fishermen on a jetty silhouetted against a dark red sunset sky. Mr. I. was totally unable to see the fishermen or the jetty, and saw only the half-engulfed hemisphere of the setting sun.

Though such problems arose when he was shown colored pictures, Mr. I. had no difficulty describing black-and-white photographs or reproductions accurately; he had no difficulty recognizing forms. His imagery and memory of objects and pictures shown to him were indeed exceptionally vivid and accurate, though always colorless. Thus, after being given a classic test picture of a colored boat, he looked intensely, looked away, and then rapidly reproduced it in black and white paint. When asked the colors of familiar objects, he had no difficulties in color association or color naming. (Patients with color anomia, for instance, can match colors perfectly but have lost the *names* of colors, and might speak, uncertainly, of a banana being "blue." A patient with a color agnosia, by contrast, could also match colors, but would evince no surprise if *given* a blue banana. Mr. I., however, had neither of these problems.)[9] Nor did he (now) have any difficulties reading. Testing up to this point, and a general neurological examination, thus confirmed Mr. I.'s total achromatopsia.

We could say to him at this point that his problem was real—that he had a true achromatopsia and not a hysteria. He took this, we thought, with mixed feelings: he had half hoped it might be merely a hysteria, and as such potentially reversible. But the notion of something psychological had also dis-

---

[9] In 1877, Gladstone, in an article entitled "On the Colour Sense of Homer," spoke of Homer's use of such phrases as "the wine-dark sea." Was this just a poetic convention, or did Homer, the Greeks, actually see the sea differently? There is indeed considerable variation between different cultures in the way they will categorize and name colors—individuals may only "see" a color (or make a perceptual categorization) if there is an existing cultural category or name for it. But it is not clear whether such categorization may actually alter elementary color perception.

tressed him and made him feel that his problem was "not real" (indeed, several doctors had hinted at this). Our testing, in a sense, legitimized his condition, but deepened his fear about brain damage and the prognosis for recovery.

Although it seemed that he had an achromatopsia of cerebral origin, we could not help wondering whether a lifetime of heavy smoking could have played a part; nicotine can cause a dimming of vision (an amblyopia) and sometimes an achromatopsia—but this is predominantly due to its effects on the cells of the retina. But the major problem was clearly cerebral: Mr. I. could have sustained tiny areas of brain damage as a result of his concussion; he could have had a small stroke either following, or conceivably precipitating, the accident.

The history of our knowledge about the brain's ability to represent color has followed a complex and zigzag course. Newton, in his famous prism experiment in 1666, showed that white light was composite—could be decomposed into, and recomposed by, all the colors of the spectrum. The rays that were bent most ("the most refrangible") were seen as violet, the least refrangible as red, with the rest of the spectrum in between. The colors of objects, Newton thought, were determined by the "copiousness" with which they reflected particular rays to the eye. Thomas Young, in 1802, feeling that there was no need to have an infinity of different receptors in the eye, each tuned to a different wavelength (artists, after all, could create almost any color they wanted by using a very limited palette of paints) postulated that three types of receptors would be enough.[10] Young's brilliant idea, thrown off ca-

---

[10] "As it is almost impossible to conceive each sensitive point of the retina to contain an infinite number of particles, each capable of vibrating in perfect unison with every possible undulation," Young wrote, "it becomes necessary to suppose the number limited, for instance to the three principal colours, red, yellow, and blue."

The great chemist John Dalton, just five years earlier, had provided a classic description of red-green colorblindness in himself. He thought this was due to a discoloration in the transparent media of the eye—and, indeed, willed an eye to

sually in the course of a lecture, was forgotten, or lay dormant, for fifty years, until Hermann von Helmholtz, in the course of his own investigation of vision, resurrected it and gave it a new precision, so that we now speak of the Young-Helmholtz hypothesis. For Helmholtz, as for Young, color was a direct expression of the wavelengths of light absorbed by each receptor, the nervous system just translating one into the other: "Red light stimulates the red-sensitive fibres strongly, and the other two weakly, giving the *sensation* red."[11]

In 1884, Hermann Wilbrand, seeing in his neurological practice patients with a range of visual losses—in some predominantly the loss of visual field, in others predominantly of color perception, and in still others predominantly of form perception—suggested that there must be separate visual centers in the primary visual cortex for "light impressions," "color impressions," and "form impressions," though he had no anatomical evidence for this. That achromatopsia (and even hemi-achromatopsia) could indeed arise from damage to specific parts of the brain was first confirmed, four years later, by a Swiss ophthalmologist, Louis Verrey. He described a sixty-year-old woman who, in consequence of a stroke affecting the occipital lobe of her left hemisphere, now saw everything in the right half of her visual field in shades of grey (the left half

---

posterity to test this. Young, however, provided the correct interpretation—that one of the three types of color receptor was missing. (Dalton's eye still resides, pickled, on a shelf in Cambridge.)

Lindsay T. Sharpe and Knut Nordby discuss this and many other aspects of the history of colorblindness research in "Total Colorblindness: An Introduction."

[11] In 1816, the young Schopenhauer proposed a different theory of color vision, one that envisaged not a passive, mechanical resonance of tuned particles or receptors, as Young had postulated, but their active stimulation, competition, and inhibition—an explicit "opponens" theory such as Ewald Hering was to create seventy years later, in apparent contradiction of the Young-Helmholtz theory. These opponens theories were ignored at the time, and continued to be ignored until the 1950s. We now envisage a combination of Young-Helmholtz and opponens mechanisms: tuned receptors, which converse with one another, are continually linked in an interactional balance. Thus integration and selection, as Schopenhauer divined, start in the retina.

remained normally colored). The opportunity to examine his patient's brain after her death showed damage confined to a small portion (the fusiform and lingual gyri) of the visual cortex—it was here, Verrey concluded, that "the centre for chromatic sense will be found." That such a center might exist, that any part of the cortex might be specialized for the perception or representation of color, was immediately contested and continued to be contested for almost a century. The grounds of this contention go very deep, as deep as the philosophy of neurology itself.

Locke, in the seventeenth century, had held to a "sensationalist" philosophy (which paralleled Newton's physicalist one): our senses are measuring instruments, recording the external world for us in terms of sensation. Hearing, seeing, all sensation, he took to be wholly passive and receptive. Neurologists in the late nineteenth century were quick to accept this philosophy and to embed it in a speculative anatomy of the brain. Visual perception was equated with "sense-data" or "impressions" transmitted from the retina to the primary visual area of the brain, in an exact, point-to-point correspondence—and there experienced, subjectively, as an image of the visual world. Color, it was presumed, was an integral part of this image. There was no room, anatomically, it was thought, for a separate color center—or indeed, conceptually, for the very idea of one. Thus when Verrey published his findings in 1888, they flew in the face of accepted doctrine. His observations were doubted, his testing criticized, his examination regarded as flawed—but the real objection, behind these, was doctrinal in nature.

If there was no discrete color center, so the thinking went, there could be no isolated achromatopsia either; thus Verrey's case, and two similar ones in the 1890s, were dismissed from neurological consciousness—and cerebral achromatopsia, as a subject, all but disappeared for the next seventy-five years.[12]

---

[12] There is no mention of it in the great 1911 edition of Helmholtz's *Physiological Optics*, though there is a large section on retinal achromatopsia.

There was not to be another full case study until 1974.[13]

Mr. I. himself was actively curious about what was going on in his brain. Though he now lived wholly in a world of lightnesses and darknesses, he was very struck by how these changed in different illuminations; red objects, for instance, which normally appeared black to him, became lighter in the long rays of the evening sun, and this allowed him to infer their redness. This phenomenon was very marked if the quality of illumination suddenly changed, as, for example, when a fluorescent light was turned on, which would cause an immediate change in the brightnesses of objects around the room. Mr. I. commented that he now found himself in an inconstant world, a world whose lights and darks fluctuated with the wavelength of illumination, in striking contrast to the relative stability, the constancy, of the color world he had previously known.[14]

All of this, of course, is very difficult to explain in terms of classical color theory—Newton's notion of an invariant relationship between wavelength and color, of a cell-to-cell transmission of wavelength information from the retina to the brain, and of a direct conversion of this information into color. Such a simple process—a neurological analogy to the decomposition and

---

[13] There were, however, brief mentions of achromatopsia in these intervening years, which were ignored, or soon forgotten, for the most part. Even Kurt Goldstein, although philosophically opposed to notions of isolated neurological deficits, remarked that he had seen several cases of pure cerebral achromatopsia without visual field losses or other impairments—an observation thrown off casually in the course of his 1948 book, *Language and Language Disturbances*.

[14] A perhaps similar phenomenon is described by Knut Nordby. During his first school year, his teacher presented the class with a printed alphabet, in which the vowels were red and the consonants black.

> I could not see any difference between them and could not understand what the teacher meant, until early one morning late in the autumn when the room-lights had been turned on, and, unexpectedly, I saw that some of the letters, i.e. the AEIOUY ÅÄÖ, were now suddenly a darkish grey, while the others were still solid black. This experience taught me that colours may look different under different light-sources, and that the same colour can be matched to different grey-tones in different kinds of illumination.

recomposition of light through a prism—could hardly account for the complexity of color perception in real life.

This incompatibility between classical color theory and reality struck Goethe in the late eighteenth century. Intensely aware of the phenomenal reality of colored shadows and colored afterimages, of the effects of contiguity and illumination on the appearance of colors, of colored and other visual illusions, he felt that these must be the basis of a color theory and declared as his credo, "Optical illusion is optical truth!" Goethe was centrally concerned with the way we actually see colors and light, the ways in which we *create* worlds, and illusions, in color. This, he felt, was not explicable by Newton's physics, but only by some as-yet unknown rules of the brain. He was saying, in effect, "Visual illusion is neurological truth."

Goethe's color theory, his *Farbenlehre* (which he regarded as the equal of his entire poetic opus), was, by and large, dismissed by all his contemporaries and has remained in a sort of limbo ever since, seen as the whimsy, the pseudoscience, of a very great poet. But science itself was not entirely insensitive to the "anomalies" that Goethe considered central, and Helmholtz, indeed, gave admiring lectures on Goethe and his science, on many occasions—the last in 1892. Helmholtz was very conscious of "color constancy"—the way in which the colors of objects are preserved, so that we can categorize them and always know what we are looking at, despite great fluctuations in the wavelength of the light illuminating them. The actual wavelengths reflected by an apple, for instance, will vary considerably depending on the illumination, but we consistently see it as red, nonetheless. This could not be, clearly, a mere translation of wavelength into color. There had to be some way, Helmholtz thought, of "discounting the illuminant"—and this he saw as an "unconscious inference" or "an act of judgement" (though he did not venture to suggest where such judgement might occur). Color constancy, for him, was a special example of the way in which we achieve perceptual constancy generally, make a stable perceptual world from a chaotic sensory flux—a world that would not be possible if our perceptions were merely passive reflections of the unpredictable and inconstant input that bathes our receptors.

Helmholtz's great contemporary, Clerk Maxwell, had also been fascinated by the mystery of color vision from his student days. He formalized the notions of primary colors and color mixing by the invention of a color top (the colors of which fused, when it was spun, to yield a sensation of grey), and a graphic representation with three axes, a color triangle, which showed how any color could be created by different mixtures of the three primary colors. These prepared the way for his most spectacular demonstration, the demonstration in 1861 that color photography was possible, despite the fact that photographic emulsions were themselves black and white. He did this by photographing a colored bow three times, through red, green, and violet filters. Having obtained three "color-separation" images, as he called them, he now brought these together by superimposing them upon a screen, projecting each image through its corresponding filter (the image taken through the red filter was projected with red light, and so on). Suddenly, the bow burst forth in full color. Maxwell wondered if this was how colors were perceived in the brain, by the addition of color-separation images or their neural correlates, as in his magic-lantern demonstrations.[15]

---

[15] Maxwell's demonstration of the "decomposition" and "reconstitution" of color in this way made color photography possible. Huge "color cameras" were used at first, which split the incident light into three beams and passed these through filters of the three primary colors (such a camera, reversed, served as a chromoscope, or Maxwellian projector). Though an integral color process was envisaged by Ducos du Hauron in the 1860s, it was not until 1907 that such a process (Autochrome) was actually developed, by the Lumière brothers. They used tiny starch grains dyed red, green, and violet, in contact with the photographic emulsion—these acted as a sort of Maxwellian grid through which the three color-separation images, mosaicked together, could both be taken and viewed. (Color cameras, Lumièrecolor, Dufaycolor, Finlaycolor, and many other additive color processes were still being used in the 1940s, when I was a boy, and stimulated my own first interest in the nature of color.)

Maxwell himself was acutely aware of the drawback of this additive process: color photography had no way of "discounting the illuminant," and its colors changed helplessly with changing wavelengths of light.

In 1957, ninety-odd years after Maxwell's famous demonstration, Edwin Land—not merely the inventor of the instant Land camera and Polaroid, but an experimenter and theorizer of genius—provided a photographic demonstration of color perception even more startling. Unlike Maxwell, he made only two black-and-white images (using a split-beam camera so they could be taken at the same time from the same viewpoint, through the same lens) and superimposed these on a screen with a double-lens projector. He used two filters to make the images: one passing longer wavelengths (a red filter), the other passing shorter wavelengths (a green filter). The first image was then projected through a red filter, the second with ordinary white light, unfiltered. One might expect that this would produce just an overall pale-pink image, but something "impossible" happened instead. The photograph of a young woman appeared instantly in full color—"blonde hair, pale blue eyes, red coat, bluegreen collar, and strikingly natural flesh tones," as Land later described it. Where did these colors come from, how were they made? They did not seem to be "in" the photographs or the illuminants themselves. These demonstrations, overwhelming in their simplicity and impact, were color "illusions" in Goethe's sense, but illusions that demonstrated a neurological truth—that colors are not "out there" in the world, nor (as classical theory held) an automatic correlate of wavelength, but, rather, are *constructed by the brain*.

These experiments hung, at first, like anomalies, conceptless, in midair; they were inexplicable in terms of existing theory, but did not yet point clearly to a new one. It seemed possible, moreover, that the viewer's knowledge of appropriate colors might influence his perception of such a scene. Land decided, therefore, to replace familiar images of the natural world with entirely abstract, multicolored displays consisting of geometric patches of colored paper, so that expectation could provide no clues as to what colors should be seen. These abstract displays vaguely resembled some of the paintings of Piet Mondrian, and Land therefore terms them "color Mondrians." Using the Mondrians, which were illuminated by three projectors, using long-wave (red), middle-wave (green), and short-wave (blue) filters, Land was able to prove that, if a surface formed part of a complex multicolored scene, there was no simple relationship between the wavelength of light reflected from a surface and its perceived color.

If, moreover, a single patch of color (for example, one ordinarily seen as green) was isolated from its surrounding colors, it would appear only as white or pale grey, whatever illuminating beam was used. Thus the green patch, Land showed, could not be regarded as inherently green, but was, in part, *given* its greenness by its relation to the surrounding areas of the Mondrian.

Whereas color for Newton, for classical theory, was something local and absolute, given by the wavelength of light reflected from each point, Land showed that its determination was neither local nor absolute, but depended upon the surveying of a whole scene and a comparison of the wavelength composition of the light reflected from each point with that of the light reflected from its surround. There had to be a continuous relating, a comparison of every part of the visual field with its own surround, to arrive at that global synthesis—Helmholtz's "act of judgement." Land felt that this computation or correlation followed fixed, formal rules; and he was able to predict which colors would be perceived by an observer under different conditions. He devised a "color cube," an algorithm, for this, in effect a model for the brain's comparison of the brightnesses, at different wavelengths, of all the parts of a complex, multicolored surface. Whereas Maxwell's color theory and color triangle were based on the concept of color addition, Land's model was now one of comparison. He proposed that there were, in fact, two comparisons: first of the reflectance of all the surfaces in a scene within a certain group of wavelengths, or waveband (in Land's term, a "lightness record" for that waveband), and second, a comparison of the three separate lightness records for the three wavebands (corresponding roughly to the red, green, and blue wavelengths). This second comparison generated the color. Land himself was at pains to avoid specifying any particular brain site for these operations and was careful to call his theory of color vision the Retinex theory, implying that there might be multiple sites of interaction between the retina and the cortex.

If Land was approaching the problem of how we see colors at a psychophysical level by asking human subjects to report how they perceived complex, multicolored mosaics in changing illuminations, Semir Zeki, working in London, was approaching the problem at a physiological level, by inserting microelectrodes in the visual cortex of anesthetized monkeys and measuring the neuronal potentials generated when they were given colored stimuli. Early in the 1970s, he was able to make a crucial discovery, to delineate a small area of cells on each side of the brain, in the prestriate cortex of monkeys (areas referred to as V4), which seemed to be specialized for responding to color (Zeki called these "color-coding cells").[16]

---

[16] He was also able to find cells, in an adjacent area, that seemed to respond solely to movement. A remarkable account and analysis of a patient with a pure "motion blindness" was given by Zihl, Von Cramon, and Mai in 1983. The patient's problems are described as follows:

The visual disorder complained of by the patient was a loss of movement vision in all three dimensions. She had difficulty, for example, in pouring tea or coffee into a cup because the fluid appeared to be frozen, like a glacier. In addition, she could not stop pouring at the right time since she was unable to perceive the movement in the cup (or a pot) when the fluid rose. Furthermore the patient complained of difficulties in following a dialogue because she could not see the movement of the face and, especially, the mouth of the speaker. In a room where more than two other people were walking, she felt very insecure and unwell, and usually left the room immediately, because "people were suddenly here or there but I have not seen them moving." The

Thus, ninety years after Wilbrand and Verrey had postulated a specific center for color in the brain, Zeki was finally able to prove that such a center existed.

Fifty years earlier, the eminent neurologist Gordon Holmes, reviewing two hundred cases of visual problems caused by gunshot wounds to the visual cortex, had found not a single case of achromatopsia. He went on to deny that an isolated cerebral achromatopsia *could* occur. The vehemence of this denial, coming from such a great authority, played a major part in bringing all clinical interest in the subject to an end.[17] Zeki's brilliant and undeniable demonstration startled the neurological world, reawakening attention to a subject it had for many years dismissed. Following his 1973 paper, new cases of human achromatopsia began appearing in the literature once again, and these could now be examined with new brain-imaging techniques (CAT, MRI, PET, SQUID, etc.) not available to neurologists of an earlier era. Now, for the first time, it was possible to visualize, in life, what areas of the brain might be needed for human color perception. Though many of the cases described had other problems, too (cuts in the visual field, visual agnosia, alexia, etc.), the crucial lesions seemed to be in the medial association cortex, in areas homologous to V4 in the monkey.[18] It had been shown in the 1960s that there

---

patient experienced the same problem but to an even more marked extent in crowded streets or places, which she therefore avoided as much as possible. She could not cross the street because of her inability to judge the speed of a car, but she could identify the car itself without difficulty. "When I'm looking at the car first, it seems far away. But then, when I want to cross the road, suddenly the car is very near." She gradually learned to "estimate" the distance of moving vehicles by means of the sound becoming louder.

[17] A vivid account of Holmes's negative influence has been provided by Damasio, who also points out that all of Holmes's cases involved lesions in the dorsal aspect of the occipital lobe, whereas the center for achromatopsia lies on the ventral aspect.

[18] The work of Antonio and Hanna Damasio and their colleagues at the University of Iowa was particularly important here, both by virtue of the minuteness of the perceptual testing, and the refinement of the neuroimaging they used.

were cells in the primary visual cortex of monkeys (in the area termed V1) that responded specifically to wavelength, but not to color; Zeki now showed, in the early 1970s, that there were other cells in the V4 areas that responded to color but not to wavelength (these V4 cells, however, received impulses from the V1 cells, converging through an intermediate structure, V2). Thus each V4 cell received information regarding a large portion of the visual field. It seemed that the two stages postulated by Land in his theory might now have an anatomical and physiological grounding: lightness records for each waveband being extracted by the wavelength-sensitive cells in V1, but only being compared or correlated to generate color in the color-coding cells of V4. Every one of these, indeed, seemed to act as a Landian correlator, or a Helmholtzian "judge."

Color vision, it seemed—like the other processes of early vision: motion, depth, and form perception—required no prior knowledge, was not determined by learning or experience, but was, as neurologists say, a "bottom-up" process. Color can indeed be generated, experimentally, by magnetic stimulation of V4, causing the "seeing" of colored rings and halos—so-called chromatophenes.[19] But color vision, in real life, is part and parcel of our total experience, is linked with our own categorizations and values, becomes for each of us a part of our life-

---

[19] Such chromatophenes may occur spontaneously in visual migraines, and Mr. I. himself had experienced these, on occasion, in migraines occurring before his accident. One wonders what would have been experienced if Mr. I.'s V4 areas had been stimulated—but magnetic stimulation of circumscribed brain areas was not technically possible at the time. One wonders, too, now that such stimulation is possible, whether it might be tried in individuals with congenital (retinal) achromatopsia (several such achromatopes have expressed their curiosity about such an experiment). It is possible—I am not aware of any studies on this—that V4 fails to develop in such people, with the absence of any cone input. But if V4 *is* present as a functional (though never functioning) unit despite the absence of cones, its stimulation might produce an astounding phenomenon—a burst of unprecedented, totally novel sensation, in a brain/mind that had never had a chance to experience or categorize such sensation. Hume wonders if a man could imagine, could even perceive, a color he had never seen before—perhaps this Humean question (propounded in 1738) could find an answer now.

world, of us. V4 may be an ultimate generator of color, but it signals to, it converses with, a hundred other systems in the mind-brain; and perhaps it can also be modulated by these. It is at higher levels that integration occurs, that color fuses with memories, expectations, associations, and desires to make a world with resonance and meaning for each of us.[20]

Mr. I. not only presented a rather "pure" case of cerebral achromatopsia (virtually uncontaminated by additional defects in the perception of form, motion, or depth), but was a highly intelligent and expert witness as well, one who was skilled at drawing and reporting what he saw. Indeed, when we first met, and he described how objects and surfaces "fluctuated" in different lights, he was, so to speak, describing the world in wavelengths, not in colors. The experience was so unlike anything he had ever experienced, so strange, so anomalous, that he could find no parallels, no metaphors, no paints or words to depict it.

When I phoned Professor Zeki to tell him of this exceptional patient, he was greatly intrigued and wondered, in particular, how Mr. I. might do with Mondrian testing, such as he and Land had used with normally sighted people and with animals. He at once arranged to come to New York to join us—Bob Wasserman, my ophthalmologist colleague; Ralph Siegel, a neurophysiologist; and myself—in a comprehensive testing of Jonathan I. No patient with achromatopsia had ever been examined in this way before.

We used a Mondrian of great complexity and brilliance, illuminated either by white light or by light filtered through narrow-band filters allowing only long wavelengths (red), in-

---

[20] The power of expectation and mental set in the perception of color is clearly shown in those with partial red-green colorblindness. Such people may not, for example, be able to spot scarlet holly berries against the dark green foliage, or the delicate salmon-pink of dawn—until these are pointed out to them. "Our poor impoverished cone cells," says a dyschromatope of my acquaintance, "need the amplification of intellect, knowledge, expectation, and attention in order to 'see' the colors that we are normally 'blind' to."

termediate wavelengths (green), or short wavelengths (blue) to pass. The intensity of the illuminating beam, in every case, was the same.

Mr. I. could distinguish most of the geometric shapes, though only as consisting of differing shades of grey, and he instantly ranked them on a one-to-four grey scale, although he could not distinguish some color boundaries (for example, between red and green, which both appeared to him, in white light, as black). With rapid, random switching of the filters, the grey-scale value of all the shapes dramatically changed— some shades previously indistinguishable now became very different, and all shades (except actual black) changed, either grossly or subtly, with the wavelength of the illuminating beam. (Thus a green area would be seen by him as white in medium-wavelength light, but as black in white or long-wavelength light.)

All Mr. I.'s responses were consistent and immediate. (It would have been very difficult, if not impossible, for a normally sighted person to make these instant and invariably "correct" estimations, even with a perfect memory and a profound knowledge of the latest color theory.) Mr. I., it was clear, *could* discriminate wavelengths, but he could not go on from this to translate the discriminated wavelengths into color; he could not generate the cerebral or mental construct of color.

This finding not only clarified the nature of the problem, but also served to pinpoint the location of the trouble. Mr. I.'s primary visual cortex was essentially intact, and it was the secondary cortex (specifically the V4 areas, or their connections) that bore virtually the whole brunt of the damage. These areas are very small, even in man; yet all our perception of color, all our ability to imagine or remember it, all our sense of living in a world of color, depend crucially on their integrity. A mischance had devastated these bean-sized areas of Mr. I.'s brain—and with this, his whole life, his life-world, had been changed.

The Mondrian testing had demonstrated damage in these

areas; we wondered now if we could see this, using brain scans. But CAT and MRI scans were entirely normal. This could have been because the scanning techniques of the time had a resolution inadequate to visualize what may have been only a patchy damage to V4; it could have been that the damage sustained was metabolic only, not structural; or it could have been that the main damage was not in V4 itself, but in the structures (the so-called "blobs" in V1 or the "stripes" in V2) leading up to it.[21]

It has been stressed—by both Zeki and Francis Crick—that these small structures, the blobs and stripes, are intensely active metabolically and may be unusually vulnerable to even temporary reductions of oxygen. Crick, in particular (with whom I discussed the case in great detail), wondered whether Mr. I. could have suffered from carbon monoxide poisoning, which is known to cause changes in color vision through its effects on the oxygenation of the blood to the color areas. Mr. I. might have been exposed to carbon monoxide through a leaky exhaust in his car, Crick speculated—perhaps due to the accident, conceivably even causing it.[22]

But all this was in a sense academic. Mr. I.'s achromatopsia, after three months, remained absolute, and he had persisting

---

[21] Malfunction in V4 can be shown by a newer technique, PET scanning (which pictures the metabolic activity of different brain areas), even if no anatomical lesion is visible on CAT or MRI scans. Unfortunately, this was not available to us at the time.

[22] Mr. I., fond of spending time in sports clubs and bars, did some research here himself and told us that he had spoken to a number of boxers who had had transient, and sometimes persistent, losses of color vision following blows to the head. Partial or total achromatopsia ("greying-out"), also temporary, is characteristic of fainting or shock, in which there is a reduction of blood supply to the posterior, and especially the visual parts, of the brain. Greying-out also occurs in transient ischemic attacks, due to arterial insufficiency—Zeki speculates that this affects the wavelength-selective cells in the blobs of V1 and the thin stripes of V2. Transient alterations of color vision—including bizarre instabilities or transformations of color (dyschromatopsia)—may also occur in visual migraines and epilepsies and are well known to users of mescaline and other drugs. They can be a disquieting side effect of ibuprofen.

impairments of contrast vision, too.[23] Whether these would clear eventually we could not say—some cases of acquired cerebral achromatopsia improve with time, but others do not. We still did not know what had caused the damage to Mr. I.'s brain, whether it was a toxin such as carbon monoxide, or the impact of the car accident, or the result of an impairment of blood flow to the visual areas of the brain. It was possible that if it had been caused by a stroke, there might be more such strokes. The prognosis had to remain uncertain, although his situation by now seemed to be stable.

We were, however, able to offer a little practical help. Mr. I. had consistently seen the boundaries of the Mondrian patches most clearly when these were illuminated by medium-wavelength light, and Dr. Zeki therefore suggested we give him a pair of green sunglasses, transmitting only this waveband in which he saw most clearly. A pair of glasses was specially made, and Mr. I. took to wearing them, especially in bright sunlight. The new glasses delighted him, for although they did nothing to restore his lost color vision, they did seem noticeably to enhance his contrast vision and his perception of form and boundaries. He could even enjoy color TV with his wife again. (The dark-green glasses, in effect, rendered the color set monochromatic—though he continued to prefer his old black-and-white set when alone.)

The sense of loss following his accident was overwhelming to Jonathan I., as it must be to anyone who loses color, a sense

---

[23] It was never quite clear from Mr. I.'s descriptions of daily life whether or not he had some slight impairment of form vision. But, interestingly, when he was being tested on the Mondrians, boundaries between rectangles tended to disappear with prolonged fixation, though they would be rapidly restored if the stimulus was moved. There are two other systems besides the blob system in early visual processing: the M system, which deals with movement and depth perception particularly, but not color; and a P-interblob system, which probably deals with high-resolution form perception. Zeki thought that the dissolution of boundaries with prolonged fixation suggested a defect in the P system, and their rapid restoration with movement "a healthy and active M system."

that interweaves itself in all our visual experiences and is so central in our imagination and memory, our knowledge of the world, our culture and art. This sense of loss, in relation to the natural world, has been remarked upon in every case. For the nineteenth-century physician thrown from his horse, flowers had "lost more than half their beauty," and entering his garden, abruptly bereft of color, was not short of shocking. This sense of loss and of shock was doubled and redoubled for Mr. I., for he had not only lost the beauty of the natural world, and the world of people, and of the innumerable objects whose colors are part of daily life, but he had also lost the world of art, he felt—the world that, for fifty years or more, had absorbed his profoundly visual and chromatic talents and sensibilities. The first weeks of his achromatopsia were thus weeks of an almost suicidal depression.[24]

In addition to his sense of loss, Jonathan I. found his

---

[24] This sense of loss is not, of course, experienced by those born totally colorblind. This is brought out in another letter I received recently from a charming and intelligent woman, Frances Futterman, born totally colorblind. She contrasted her own situation with that of Jonathan I.:

> I was struck by how different that kind of experience must be, compared to my own experience of never having seen color before, thus never having lost it—and also never having been depressed about my colorless world. . . . The way I see in and of itself is not depressing. In fact, I am frequently overwhelmed by the beauty of the natural world. . . . People say I must see in shades of gray or in "black and white," but I don't think so. The word gray has no more meaning for me than the word pink or blue—in fact, even less meaning, because I have developed inner concepts of color words like pink and blue; but, for the life of me, I can't conceive of gray.

Though Mrs. Futterman's experience is certainly different from Mr. I.'s, both remark on the meaninglessness of the word "grey," a word that can no more convey anything to the achromatopic than can "darkness" to the blind, or "silence" to the deaf. Mrs. Futterman remarks, as Mr. I. came to, on the beauty of her world. "I would also be willing to bet," she says, "that if we were tested along with normals in low lighting levels, we would be able to detect far more shades of gray. Black and white photos look far too stark to me. The world I see has so much more richness and variety than black-and-white photos or TV shows. . . . My vision is a lot richer than normals can imagine."

changed visual world, at first, abhorrent and abnormal. This, too, is the experience of most people in his position: the concussed physician thrown from his horse found his vision "perverted," one of Damasio's patients found her grey world "dirty." Why, one must wonder, do all subjects with a cerebral achromatopsia express themselves in such terms—why should their experience seem so abnormal? Mr. I. was seeing with his cones, seeing with the wavelength-sensitive cells of V1, but unable to use the higher-order, color-generating mechanism of V4. For us, the output of V1 is unimaginable, because it is never experienced as such and is immediately shunted on to a higher level, where it is further processed to yield the perception of color. Thus the raw output of V1 never appears in awareness for us. But for Mr. I. it did—his brain damage had made him privy to, indeed trapped him within, a strange in-between state—the uncanny world of V1—a world of anomalous and, so to speak, prechromatic sensation, which could not be categorized as either colored *or* colorless.[25]

Mr. I., with his heightened visual and aesthetic sensibilities, found these changes particularly intolerable. We know too little about what determines emotion and aesthetic appeal in relation to color, and indeed in relation to seeing generally—and this is a matter of individual experience and taste.[26]

Color perception had been an essential part not only of

---

[25] We may experience something like this, Zeki has recently shown, by using an inhibitory magnetic stimulation to V4, which produces a temporary achromatopsia.

[26] We also know too little about the interactions of the three major systems in early vision—the M, interblob, and blob systems. But Crick wonders whether some of the unpleasantness and abnormality, at least—the "leaden" vision of which Mr. I. complained—might not in part be due to the unmoderated action of the preserved M system, which, he emphasizes, "sees few shades of grey, [so that] its white would correspond to what was (in normal people) a dirty white." This notion gains support from the fact that people with *congenital* achromatopsia, who have not sustained any damage to their higher visual systems, do not have any such perceptual abnormalities. Thus Knut Nordby writes: "I have never experienced 'dirty,' 'impure,' 'stained,' or 'washed-out' colors, as reported by the artist Jonathan I."

Mr. I.'s visual sense, but his aesthetic sense, his sensibility, his creative identity, an essential part of the way he constructed his world—and now color was gone, not only in perception, but in imagination and memory as well. The resonances of this were very deep. At first he was intensely, furiously conscious of what he had lost (though "conscious," so to speak, in the manner of an amnesiac). He would glare at an orange in a state of rage, trying to force it to resume its true color. He would sit for hours before his (to him) dark grey lawn, trying to see it, to imagine it, to remember it, as green. He found himself now not only in an impoverished world, but in an alien, incoherent, and almost nightmarish one. He expressed this soon after his injury, better than he could in words, in some of his early, desperate paintings.

But then, with the "apocalyptic" sunrise, and his painting of this, came the first hint of a change, an impulse to construct the world anew, to construct his own sensibility and identity anew. Some of this was conscious and deliberate: retraining his eyes (and hands) to operate, as he had in his first days as an artist. But much occurred below this level, at a level of neural processing not directly accessible to consciousness or control. In this sense, he started to be redefined by what had happened to him—redefined physiologically, psychologically, aesthetically—and with this there came a transformation of values, so that the total otherness, the alienness of his V1 world, which at first had such a quality of horror and nightmare, came to take on, for him, a strange fascination and beauty.

Immediately after his accident, and for a year or more thereafter, Jonathan I. insisted that he still "knew" colors, knew what was right, what was appropriate, what was beautiful, even if he could no longer visualize them in his mind. But, thereafter, he became somewhat less sure, as if now, unsupported by actual experience or image, his color associations had started to give way. Perhaps such a forgetting—a forgetting at once physiological and psychological, at once strategic and structural—may have to occur, to some extent, sooner or later, in anyone who is no longer able to experience or imagine, or in any way to generate, a particular mode of perception. (Nor is it necessary that the primary damage be cortical; it may occur, after months or years, even in those who are peripherally or retinally blind.)[27]

There was a lessening concern with what he had lost, and indeed with the whole subject of color, which at first had so obsessed him. Indeed, he now spoke of being "divorced" from color. He could still speak fluently about it, but there seemed to be a certain hollowness to his words, as if he were drawing only from past knowledge and no longer *understood* it.

Nordby writes:

> Although I have acquired a thorough theoretical knowledge of the physics of colours and the physiology of the colour receptor mechanisms, nothing of this can help me to understand the true nature of colours.[28]

What was true for Nordby was now true for Jonathan I., too. He had in some ways started to resemble a person born colorblind, even though he had lived in a color world for the first sixty-five years of his life.

At once forgetting and turning away from color, turning

---

[27] J. D. Mollon et al. describe the case of a young police cadet who, following a severe febrile illness (probably cerebral herpes) was left with achromatopsia, hemianopia, and some agnosia and amnesia. Testing him five years after the illness, Mollon reports that "he was able to name (presumably by means of verbal memory) the colours of e.g., grass, traffic lights, and the union jack, but made errors on other common objects (e.g., banana, pillar-box)." Thus here, after five years of total colorblindness, the colors of even the most familiar objects were often forgotten. Such effects have been recorded in ordinary retinal blindness, too, where after some years there may be a widespread loss of all visual memories, including those of color.

[28] "A very intelligent blind person," Schopenhauer writes, "could almost [construct] a theory of colours from accurate statements that he heard about them." Diderot, similarly, speaking of Nicholas Saunderson, a famous blind lecturer on optics at Oxford in the early eighteenth century, feels that he had a profound theoretical knowledge and *concept* of space, although he never had any direct visual *percept* of it. (See footnote 13, page 139.)

away from the chromatic orientation and habits and strategies of his previous life, Mr. I., in the second year after his injury, found that he saw best in subdued light or twilight, and not in the full glare of day. Very bright light tended to dazzle and temporarily blind him—another sign of damage to his visual systems—but he found the night and nightlife peculiarly congenial, for they seemed to be "designed," as he once said, "in terms of black and white."

He started becoming a "night person," in his own words, and took to exploring other cities, other places, but only at night. He would drive, at random, to Boston or Baltimore, or to small towns and villages, arriving at dusk, and then wandering about the streets for half the night, occasionally talking to a fellow walker, occasionally going into little diners: "Everything in diners is different at night, at least if it has windows. The darkness comes into the place, and no amount of light can change it. They are transformed into night places. I love the nighttime," Mr. I. said. "Gradually I am becoming a night person. It's a different world: there's a lot of space— you're not hemmed in by streets, by people. . . . It's a whole new world."

Mr. I., when he was not traveling, would get up earlier and earlier, to work in the night, to relish the night. He felt that in the night world (as he called it) he was the equal, or the superior, of "normal" people: "I feel better because I know then that I'm not a freak . . . and I have developed acute night vision, it's amazing what I see—I can read license plates at night from four blocks away. You couldn't see it from a block away."[29]

---

[29] With his revulsion from color and brightness, his fondness of dusk and night, his apparently enhanced vision at dusk and night, Mr. I. sounds like Kaspar Hauser, the boy who was confined in a dimly lit cellar for fifteen years, as Anselm von Feuerbach described him in 1832:

As to his sight, there existed, in respect to him, no twilight, no night, no darkness. . . . At night he stepped everywhere with the greatest confidence; and in dark places, he always refused a light when it was offered to him. He often

One wonders whether his night vision might, with time, have taken on heightened function in compensation for the damage to his color system—there might, at this stage, also have been a heightening of movement sensitivity, perhaps of depth sensitivity, too, possibly going with an increased dependence on and use of the intact M system.[30]

Most interesting of all, the sense of profound loss, and the sense of unpleasantness and abnormality, so severe in the first months following his head injury, seemed to disappear, or even reverse. Although Mr. I. does not deny his loss, and at some level still mourns it, he has come to feel that his vision has become "highly refined," "privileged," that he sees a world of pure form, uncluttered by color. Subtle textures and patterns, normally obscured for the rest of us because of their embedding in color, now stand out for him.[31] He feels he has been given "a whole new world," which the rest of us, dis-

---

looked with astonishment, or laughed, at persons who, in dark places, for instance, when entering a house, or walking on a staircase by night, sought safety in groping their way, or in laying hold on adjacent objects. In twilight, he even saw much better than in broad daylight. Thus, after sunset, he once read the number of a house at a distance of one hundred and eighty paces, which, in daylight, he would not have been able to distinguish so far off. Towards the close of twilight, he once pointed out to his instructor a gnat that was hanging in a very distant spider's web. (pp. 83–4)

[30] It may be that individuals with congenital achromatopsia develop heightened function of the M system, and may be extraordinarily adept at spotting movement. This is currently being investigated by Ralph Siegel and Martin Gizzi.

[31] I recently heard of an achromatopic botanist in England said to be even better than color normals at swiftly identifying ferns and other plants in woods, hedgerows, and other almost monochromatic environments. Similarly, in World War II, people with severe red-green colorblindness were pressed into service as bombardiers, because of their ability to "see through" colored camouflage and not be distracted by what would be, to the normally sighted, a confusing and deceiving configuration of colors. One veteran of the Pacific theater reports that colorblind soldiers were indispensable in spotting the movement of camouflaged troops in the jungle. (All of these things may also be clearer to color normals at twilight.)

tracted by color, are insensitive to. He no longer thinks of color, pines for it, grieves its loss. He has almost come to see his achromatopsia as a strange gift, one that has ushered him into a new state of sensibility and being. In this his transformation is exceedingly similar to that of John Hull, who, after two or three years of experiencing blindness as an affliction and curse, came to see it as "a dark, paradoxical gift," a "concentrated human condition . . . one of the orders of human being."

Once, about three years after his injury, an intriguing suggestion was made (by Israel Rosenfield), that Mr. I. try to regain his color vision. Since the mechanism for comparing wavelengths was intact, and only V4 (or its equivalent) was damaged, it might be possible, at least in theory, Rosenfield thought, to "retrain" another part of the brain to perform the requisite Landian correlations, and thus to achieve some restoration of color vision. What was striking was Mr. I.'s response to this suggestion. In the first months after his injury, he said, he would have embraced such a suggestion, done everything possible to be "cured." But now that he conceived the world in different terms, and again found it coherent and complete, he thought the suggestion unintelligible, and repugnant. Now that color had lost its former associations, its sense, he could no longer imagine what its restoration would be like. Its reintroduction would be grossly confusing, he thought, might force a welter of irrelevant sensations upon him, and disrupt the now-reestablished visual order of his world. He had been for a while in a sort of limbo; now he had settled—neurologically and psychologically—for the world of achromatopia.

In terms of his painting, after a year or more of experiment and uncertainty, Mr. I. has moved into a strong and productive phase, as strong and productive as anything in his long artistic career. His black-and-white paintings are highly successful, and people comment on his creative renewal, the remarkable black-and-white "phase" he has moved into. Very few of them know that his latest phase is anything other than an expression of his artistic development, that it was brought about by a calamitous loss.

Though it has been possible to define the primary damage in Mr. I.'s brain—the knocking out of an essential part of his color-constructing system—we are still totally ignorant of the "higher" changes in brain function that must have occurred in its train. Jonathan I. did not lose just his perception of color, but imagery, and even dreaming in color. Finally he seemed to lose even his memory of color, so that it ceased to be part of his mental knowledge, his mind.

Thus, as more and more time elapsed without color vision, he came to resemble someone with an amnesia for color—or, indeed, someone who had never known it at all. But, at the same time, a revision was occurring, so that as his former color world and even the memory of it became fainter and died inside him, a whole new world of seeing, of imagination, of sensibility, was born.[32]

There is no doubt of the reality of these changes—although it may have required a subject as gifted and as articulate as Jonathan I. to bring them out with such clarity. Neuroscience, at this point, can say nothing about the cerebral basis of such "higher" changes. The physiological investigation of color, thus far, has terminated in the color systems of early vision, the Landian correlations that occur in $V1$ and $V4$. But $V4$ is not an end point, it is only a way station, projecting in its turn to higher and higher levels—eventually to the hippocampus, so essential for the storage of memories; to the emotional centers of the limbic system and amygdala; and to many other parts of the cortex. The cessation of information flow from $V4$

---

[32] A similar emergence of new sensibilities and imagination is described in H. G. Wells's great short story "The Country of the Blind": "For fourteen generations these people have been blind and cut off from all the seeing world; the names for all the things of sight had faded and changed. . . . Much of their imagination had shrivelled with their eyes, and they had made for themselves new imaginations with their ever more sensitive ears and fingertips."

to the memory systems of the hippocampus and prefrontal cortex, for example, might in part explain Mr. I.'s "forgetting" of color. We do not have the tools at the moment to map the subtle, higher-level neural consequences of such a sensory loss, but a history such as Jonathan I.'s shows how crucial it is to do this.

Work in the last decade has shown how plastic the cerebral cortex is, and how the cerebral "mapping" of body image, for example, may be drastically reorganized and revised, not only following injuries or immobilizations, but in consequence of the special use or disuse of individual parts. We know, for instance, that the constant use of one finger in reading Braille leads to a huge hypertrophy of that finger's representation in the cortex. And with early deafness and the use of sign language, there may be drastic remappings in the brain, large areas of the auditory cortex being reallocated for visual processing. Similarly, it seemed, with Mr. I.: if entire systems of representation, of meaning, had been extinguished inside him, entirely new systems had been brought into being.

On the ultimate question—the question of qualia: why a particular sensation may be perceived as red—the case of Jonathan I. may not be able to help us at all. After describing "the celebrated phaenomenon of colours," Newton drew back from all speculation about sensation and would hazard no hypothesis as to "by what modes or action light produceth in our minds the phantasms of colours." Three centuries later, we still have no hypothesis, and perhaps such questions can never be answered at all.

The
Last Hippie

Such a long, long time to be gone . . .
and a short time to be there
> —Robert Hunter
> "Box of Rain"

Greg F. grew up in the 1950s in a comfortable Queens household, an attractive and rather gifted boy who seemed destined, like his father, for a professional career—perhaps a career in songwriting, for which he showed a precocious talent. But he grew restive, started questioning things, as a teenager in the late sixties; started to hate the conventional life of his parents and neighbors and the cynical, bellicose administration of the country. His need to rebel, but equally to find an ideal and a guide, to find a leader, crystallized in the Summer of Love, in 1967. He would go to the Village and listen to Allen Ginsberg declaiming all night; he loved rock music, especially acid rock, and, above all, the Grateful Dead.

Increasingly he fell out with his parents and teachers; he was truculent with the one, secretive with the other. In 1968, a time when Timothy Leary was urging American youth to "tune in, turn on, and drop out," Greg grew his hair long and dropped out of school, where he had been a good student; he left home and went to live in the Village, where he dropped acid and joined the East Village drug culture—searching, like others of his generation, for utopia, for inner freedom, and for "higher consciousness."

But "turning on" did not satisfy Greg, who stood in need of a more codified doctrine and way of life. In 1969 he gravitated, as so many young acidheads did, to the Swami Bhaktivedanta and his International Society for Krishna Consciousness, on Second Avenue. And under his influence, Greg, like so many others, stopped taking acid, finding his religious exaltation a replacement for acid highs. ("The only radical remedy for dipsomania," William James once said, "is religiomania.") The philosophy, the fellowship, the chanting, the rituals, the austere and charismatic figure of the swami himself, came like a revelation to Greg, and he became, almost immediately, a passionate devotee and convert.[1] Now there was a center, a focus, to his life. In those first exalted weeks of his conversion, he wandered around the East Village, dressed in saffron robes, chanting the Hare Krishna mantras, and early in 1970, he took up residence in the main temple in Brooklyn. His parents objected at first, then went along with this. "Perhaps it will help him," his father said, philosophically. "Perhaps—who knows?—this is the path he needs to follow."

Greg's first year at the temple went well; he was obedient, ingenuous, devoted, and pious. He is a Holy One, said the swami, one of us. Early in 1971, now deeply committed, Greg was sent to the temple in New Orleans. His parents had seen him occasionally when he was in the Brooklyn temple, but now communication from him virtually ceased.

One problem arose in Greg's second year with the Krishnas—he complained that his vision was growing dim, but this was interpreted, by his swami and others, in a spiritual way: he was "an illuminate," they told him; it was the "inner light" growing. Greg had worried at first about his eyesight, but was reassured by the swami's spiritual explanation. His

---

[1] The swami's unusual views are presented, in summary form, in *Easy Journey to Other Planets*, by Tridandi Goswami A. C. Bhaktivedanta Swami, published by the League of Devotees, Vrindaban (no date, one rupee). This slim manual, in its green paper cover, was handed out in vast quantities by the swami's saffron-robed followers, and it became Greg's bible at this stage.

sight grew still dimmer, but he offered no further complaints. And indeed, he seemed to be becoming more spiritual by the day—an amazing new serenity had taken hold of him. He no longer showed his previous impatience or appetites, and he was sometimes found in a sort of daze, with a strange (some said "transcendental") smile on his face. It is beatitude, said his swami—he is becoming a saint. The temple felt he needed to be protected at this stage: he no longer went out or did anything unaccompanied, and contact with the outside world was strongly discouraged.

Although Greg's parents did not have any direct communication from him, they did get occasional reports from the temple—reports filled, increasingly, with accounts of his "spiritual progress," his "enlightenment," accounts at once so vague and so out of character with the Greg they knew that, by degrees, they became alarmed. Once they wrote directly to the swami and received a soothing, reassuring reply.

Three more years passed before Greg's parents decided they had to see for themselves. His father was by then in poor health and feared that if he waited longer he might never see his "lost" son again. On hearing this, the temple finally permitted a visit from Greg's parents. In 1975, then, not having seen him for four years, they visited their son in the temple in New Orleans.

When they did so, they were filled with horror: their lean, hairy son had become fat and hairless; he wore a continual "stupid" smile on his face (this at least was his father's word for it); he kept bursting into bits of song and verse and making "idiotic" comments, while showing little deep emotion of any kind ("like he was scooped out, hollow inside," his father said); he had lost interest in everything current; he was disoriented—and he was totally blind. The temple, surprisingly, acceded to his leaving—perhaps even they felt now that his ascension had gone too far and had started to feel some disquiet about his state.

Greg was admitted to the hospital, examined, and transferred to neurosurgery. Brain imaging had shown an enormous midline tumor, destroying the pituitary gland and the adjacent optic chiasm and tracts and extending on both sides into the frontal lobes. It also reached backward to the temporal lobes, and downward to the diencephalon, or forebrain. At surgery, the tumor was found to be benign, a meningioma—but it had swollen to the size of a small grapefruit or orange, and though the surgeons were able to remove it almost entirely, they could not undo the damage it had already done.

Greg was now not only blind, but gravely disabled neurologically and mentally—a disaster that could have been prevented entirely had his first complaints of dimming vision been heeded, and had medical sense, and even common sense, been allowed to judge his state. Since, tragically, no recovery could be expected, or very little, Greg was admitted to Williamsbridge, a hospital for the chronically sick, a twenty-five-year-old boy for whom active life had come to an end, and for whom the prognosis was considered hopeless.

I first met Greg in April 1977, when he arrived at Williamsbridge Hospital. Lacking facial hair, and childlike in manner, he seemed younger than his twenty-five years. He was fat, Buddha-like, with a vacant, bland face, his blind eyes roving at random in their orbits, while he sat motionless in his wheelchair. If he lacked spontaneity and initiated no exchanges, he responded promptly and appropriately when I spoke to him, though odd words would sometimes catch his fancy and give rise to associative tangents or snatches of song and rhyme. Between questions, if the time was not filled, there tended to be a deepening silence; though if this lasted for more than a minute, he might fall into Hare Krishna chants or a soft muttering of mantras. He was still, he said, "a total believer," devoted to the group's doctrines and aims.

I could not get any consecutive history from him—he was not sure, for a start, why he was in the hospital and gave different reasons when I asked him about this; first he said, "Because I'm not intelligent," later, "Because I took drugs in the past." He knew he had been at the main Hare Krishna temple ("a big red house, 439 Henry Street, in Brooklyn"), but not that he had subsequently been at their temple in New Orleans. Nor did he remember that he started to have symptoms there—first and foremost a progressive loss of vision. Indeed he seemed unaware that he had any problems: that he was blind, that he was unable to walk steadily, that he was in any way ill.

Unaware—and indifferent. He seemed bland, placid, emptied of all feeling—it was this unnatural serenity that his Krishna brethren had perceived, apparently, as "bliss," and indeed, at one point, Greg used the term himself. "How do you feel?" I returned to this again and again. "I feel blissful," he replied at one point, "I am afraid of falling back into the material world." At this point, when he was first in the hospital, many of his Hare Krishna friends would come to visit him; I often saw their saffron robes in the corridors. They would come to visit poor, blind, blank Greg and flock around him; they saw him as having achieved "detachment," as an Enlightened One.

Questioning him about current events and people, I found the depths of his disorientation and confusion. When I asked him who was the president, he said "Lyndon," then, "the one who got shot." I prompted, "Jimmy . . . ," and he said, "Jimi Hendrix," and when I roared with laughter, he said maybe a musical White House would be a good idea. A few more questions convinced me that Greg had virtually no memory of events much past 1970, certainly no coherent, chronological memory of them. He seemed to have been left, marooned, in the sixties—his memory, his development, his inner life since then had come to a stop.

His tumor, a slow-growing one, was huge when it was finally removed in 1976, but only in the later stages of its growth, as it destroyed the memory system in the temporal lobe, would it actually have prevented the brain from registering new events. But Greg had difficulties—not absolute, but partial—even in remembering events from the late sixties, events that he must have registered perfectly at the time. So beyond the inability to register new experiences, there had been an erosion of existing memories (a retrograde amnesia) going back several years before his tumor had developed. There was not an absolutely sharp cutoff here, but rather a temporal gradient, so that figures and events from 1966 and 1967 were fully remembered, events from 1968 or 1969 partially or occasionally remembered, and events after 1970 almost never remembered.

It was easy to demonstrate the severity of his immediate amnesia. If I gave him lists of words, he was unable to recall any of them after a minute. When I told him a story and asked him to repeat it, he did so in a more and more confused way, with more and more "contaminations" and misassociations— some droll, some extremely bizarre—until within five minutes his story bore no resemblance to the one I had told him. Thus when I told him a tale about a lion and a mouse, he soon departed from the original story and had the mouse threatening to eat the lion—it had become a giant mouse and a mini-lion. Both were mutants, Greg explained when I quizzed him on his departures. Or possibly, he said, they were creatures from a dream, or "an alternative history" in which mice were indeed the lords of the jungle. Five minutes later, he had no memory of the story whatever.

I had heard, from the hospital social worker, that he had a passion for music, especially for rock-and-roll bands of the sixties; I saw piles of records as soon as I entered his room and a guitar lying against his bed. So now I asked him about this, and with this there came a complete transformation—he lost his disconnectedness, his indifference, and spoke with great animation about his favorite rock bands and pieces—above all, of the Grateful Dead. "I went to see them at the Fillmore East, and in Central Park," he said. He remembered the entire program in detail, but "my favorite," he added, "is 'Tobacco Road.' " The title evoked the tune, and Greg sang the whole song with great feeling and conviction—a depth of feeling of which, hitherto, he had not shown the least sign. He seemed transformed, a different person, a whole person, as he sang.

"When did you hear them in Central Park?" I asked.

"It's been a while, over a year maybe," he answered—but in fact they had last played there eight years earlier, in 1969. And the Fillmore East, the famous rock-and-roll theater where Greg had also seen the group, did not survive the early 1970s. He went on to tell me he once heard Jimi Hendrix at Hunter College, and Cream, with Jack Bruce playing bass guitar; Eric Clapton, lead guitar; and Ginger Baker, a "fantastic drummer." "Jimi Hendrix," he added reflectively, "what's he doing? Don't hear much about him nowadays." We spoke of the Rolling Stones and the Beatles—"Great groups," Greg commented, "but they don't space me out the way the Dead do. What a group," he continued, "there's no one like them. Jerry Garcia—he's a saint, he's a guru, he's a genius. Mickey Hart, Bill Kreutzmann, the drummers are great. There's Bob Weir, there's Phil Lesh; but Pigpen—I love him."

This narrowed down the extent of his amnesia. He remembered songs vividly from 1964 to 1968. He remembered all the founding members of the Grateful Dead, from 1967. But he was unaware that Pigpen, Jimi Hendrix, and Janis Joplin were all dead. His memory cut off by 1970, or before. He was caught in the sixties, unable to move on. He was a fossil, the last hippie.

At first I did not want to confront Greg with the enormity of his time loss, his amnesia, or even to let involuntary hints through (which he would certainly pick up, for he was very sensitive to anomaly and tone), so I changed the subject and said, "Let me examine you."

He was, I noted, somewhat weak and spastic in all his limbs, more on the left, and more in the legs. He could not stand alone. His eyes showed complete optic atrophy—it was impossible for him to see anything. But strangely, he did not seem to be *aware* of being blind and would guess that I was showing him a blue ball, a red pen (when in fact it was a green comb and a fob watch that I showed him). Nor indeed did he seem to "look"; he made no special effort to turn in my direction, and when we were speaking, he often failed to face me, to look at me. When I asked him about seeing, he acknowledged that his eyes weren't "all that good," but added that he enjoyed "watching" the TV. Watching TV for him, I observed later, consisted of following with attention the soundtrack of a movie or show and inventing visual scenes to go with it (even though he might not even be looking toward the TV). He seemed to think, indeed, that this was what "seeing" meant, that this was what was meant by "watching TV," and that this was what all of us did. Perhaps he had lost the very idea of seeing.

I found this aspect of Greg's blindness, his singular blindness to his blindness, his no longer knowing what "seeing" or "looking" meant, deeply perplexing. It seemed to point to something stranger, and more complex, than a mere "deficit," to point, rather, to some radical alteration within him in the very structure of knowledge, in consciousness, in identity itself.[2]

I had already had some sense of this when testing his memory, finding his confinement, in effect, to a single moment— "the present"—uninformed by any sense of a past (or a future). Given this radical lack of connection and continuity in his inner life, I got the feeling, indeed, that he might not *have* an in-

---

[2] Another patient, Ruby G., was in some ways similar to Greg. She too had a huge frontal tumor, which, though it was removed in 1973, left her with amnesia, a frontal lobe syndrome, and blindness. She too did not know that she was blind, and when I held up my hand before her and asked, "How many fingers?" would answer, "A hand has five fingers, of course."

A more localized unawareness of blindness may arise if there is destruction of the visual cortex, as in Anton's syndrome. Such patients may not know that they are blind, but are otherwise intact. But frontal lobe unawarenesses are far more global in nature—thus Greg and Ruby were not only unaware of being blind but unaware (for the most part) of being ill, of having devastating neurological and cognitive deficits, and of their tragic, diminished position in life.

ner life to speak of, that he lacked the constant dialogue of past and present, of experience and meaning, which constitutes consciousness and inner life for the rest of us. He seemed to have no sense of "next" and to lack that eager and anxious tension of anticipation, of intention, that normally drives us through life.

Some sense of ongoing, of "next," is always with us. But this sense of movement, of happening, Greg lacked; he seemed immured, without knowing it, in a motionless, timeless moment. And whereas for the rest of us the present is given its meaning and depth by the past (hence it becomes the "remembered present," in Gerald Edelman's term), as well as being given potential and tension by the future, for Greg it was flat and (in its meager way) complete. This living-in-the-moment, which was so manifestly pathological, had been perceived in the temple as an achievement of higher consciousness.

Greg seemed to adjust to Williamsbridge with remarkable ease, considering he was a young man being placed, probably forever, in a hospital for the chronically ill. There was no furious defiance, no railing at Fate, no sense, apparently, of indignity or despair. Compliantly, indifferently, Greg let himself be put away in the backwater of Williamsbridge. When I asked him about this, he said, "I have no choice." And this, as he said it, seemed wise and true. Indeed, he seemed eminently philosophical about it. But it was a philosophicalness made possible by his indifference, his brain damage.

His parents, so estranged from him when he was rebellious and well, came daily, doted on him, now that he was helpless and ill; and they, for their part, could be sure, at any time, that he would be at the hospital, smiling and grateful for their visit. If he was not "waiting" for them, so much the better— they could miss a day, or a few days, if they were away; he would not notice, but would be cordial as ever the next time they came.

Greg soon settled in, with his rock records and his guitar, his Hare Krishna beads, his Talking Books, and a schedule of programs—physiotherapy, occupational therapy, music groups, drama. Soon after admission he was moved to a ward with younger patients, where with his open and sunny personality he became popular. He did not actually know any of the other patients or the staff, at least for several months, but was invariably (if indiscriminately) pleasant to them all. And there were at least two special friendships, not intense, but with a sort of complete acceptance and stability. His mother remembers "Eddie, who had MS . . . they both loved music, they had adjacent rooms, they used to sit together . . . and Judy, she had CP, she would sit for hours with him, too." Eddie died, and Judy went to a hospital in Brooklyn; there has been no one so close for many years. Mrs. F. remembers them, but Greg does not, never asked for them, or about them, after they had gone—though perhaps, his mother thought, he was sadder, at least less lively, for they stimulated him, got him talking and listening to records and inventing limericks, joking and singing; they pulled him out of "that dead state" he would otherwise fall into.

A hospital for the chronically ill, where patients and staff live together for years, is a little like a village or a small town: everybody gets to meet, to know, everybody else. I often saw Greg in the corridors, being wheeled to different programs or out to the patio, in his wheelchair, with the same odd, blind yet searching look on his face. And he gradually got to know me, at least sufficiently to know my name, to ask each time we met, "How're you doing, Dr. Sacks? When's the next book coming out?" (a question that rather distressed me in the seemingly endless eleven-year interim between the publication of *Awakenings* and *A Leg to Stand On*).

Names, then, he might learn, with frequent contact, and in relation to them he would recollect a few details about each new person. Thus he came to know Connie Tomaino, the music therapist—he would recognize her voice, her footfalls, immediately—but he could never remember where or how he had met her. One day Greg began talking about "another Connie," a girl called Connie whom he'd known in high school.

This other Connie, he told us, was also, remarkably, very musical—"How come all you Connies are so musical?" he teased. The other Connie would conduct music groups, he said, would give out song sheets, play the piano-accordion at singsongs at school. At this point, it started to dawn on us that this "other" Connie was in fact Connie herself, and this was clinched when he added, "You know, she played the trumpet, too." (Connie Tomaino is a professional trumpet player.) This sort of thing often happened with Greg when he put things into the wrong context or failed to connect them with the present.

His sense of there being two Connies, his segmenting Connie into two, was characteristic of the bewilderments he sometimes found himself in, his need to hypothesize additional figures because he could not retain or conceive of an identity in time. With consistent repetition Greg might learn a few facts, and these would be retained. But the facts were isolated, denuded of context. A person, a voice, a place, would slowly become "familiar," but he remained unable to remember where he had met the person, heard the voice, seen the place. Specifically, it was context-bound (or "episodic") memory that was so grossly disturbed in Greg—as is the case with most amnesiacs.

Other sorts of memory were intact; thus Greg had no difficulty remembering or applying geometric truths that he had learned in school. He saw instantly, for example, that the hypotenuse of a triangle was shorter than the sum of the two sides—thus his semantic memory, so-called, was fairly intact. Again, he not only retained his power to play the guitar, but actually enlarged his musical repertoire, learning new techniques and fingering with Connie; he also learned to type while at Williamsbridge—so his procedural memory was also unimpaired.

Finally, there seemed to be some sort of slow habituation or familiarization—so that he became able, within three months, to find his way about the hospital, to go to the coffee shop, the cinema, the auditorium, the patio, his favorite places. This sort of learning was exceedingly slow, but once it had been achieved, it was tenaciously retained.

It was clear that Greg's tumor had caused damage that was complex and curious. First, it had compressed or destroyed structures of the inner, or medial, side of both the temporal lobes—in particular, the hippocampus and its adjacent cortex, areas crucial for the capacity to form new memories. With such damage, the ability to acquire information about new facts and events is devastated—there ceases to be any explicit or conscious remembrance of these. But while Greg was so often unable to recall events or encounters or facts to consciousness, he might nonetheless have an unconscious or implicit memory of them, a memory expressed in performance or behavior. Such implicit ability to remember allowed him to become slowly familiar with the physical layout and routines of the hospital and with some of the staff, and to make judgments on whether certain persons (or situations) were pleasant or unpleasant.[3]

While explicit learning requires the integrity of the medial temporal lobe systems, implicit learning may employ more primitive and diffuse paths, as do the simple processes of conditioning and habituation. Explicit learning, however, involves the construction of complex percepts—syntheses of representations from every part of the cerebral cortex—brought together into a contextual unity, or "scene." Such syntheses can be held in mind for only a minute or two—the limit of short-term memory—and after this will be lost unless they can be shunted into long-term memory. Thus higher-order memorization is a multistage process, involving the transfer of perceptions, or perceptual syntheses, from short-

---

[3] That implicit memory (especially if emotionally charged) may exist in amnesiacs was shown, somewhat cruelly, in 1911, by Edouard Claparède, who, when shaking hands with such a patient whom he was presenting to his students, stuck a pin in his hand. Although the patient had no explicit memory of this, he refused, thereafter, to shake hands with him.

term to long-term memory. It is just such a transfer that fails to occur in people with temporal lobe damage. Thus Greg can repeat a complicated sentence with complete accuracy and understanding the moment he hears it, but within three minutes, or sooner if he is distracted for an instant, he will retain not a trace of it, or any idea of its sense, or any memory that it ever existed.

Larry Squire, a neuropsychologist at the University of California, San Diego, who has been a central figure in elucidating this shunting function of the temporal lobe memory system, speaks of the brevity, the precariousness, of short-term memory in us all; all of us, on occasion, suddenly lose a perception or an image or a thought we had vividly in mind ("Damn it," we may say, "I've forgotten what I wanted to say!"), but only in amnesiacs is this precariousness realized to the full.

Yet while Greg, no longer capable of transforming his perceptions or immediate memories into permanent ones, remains stuck in the sixties, when his ability to learn new information broke down, he has nevertheless adapted somehow and absorbed some of his surroundings, albeit very slowly and incompletely.[4]

Some amnesiacs (like Jimmie, the patient with Korsakov's syndrome whom I described in "The Lost Mariner") have brain damage largely confined to the memory systems of the diencephalon and medial temporal lobe; others (like Mr. Thompson, described in "A Matter of Identity") are not only amnesiac but have frontal lobe syndromes, too; yet others— like Greg, with immense tumors—tend to have a third area of damage as well, deep below the cerebral cortex, in the forebrain, or diencephalon. In Greg, this widespread damage had created a very complicated clinical picture, with sometimes overlapping or even contradictory symptoms and syndromes.

---

[4] A. R. Luria, in *The Neuropsychology of Memory*, remarks that all his amnesiac patients, if hospitalized for any length of time, acquired "a sense of familiarity" with their surroundings.

Thus though his amnesia was chiefly caused by damage to the temporal lobe systems, damage to the diencephalon and frontal lobes also played a part. Similarly there were multiple origins for his blandness and indifference, for which damage to the frontal lobes, diencephalon, and pituitary gland was in varying degrees responsible. In fact, Greg's tumor first caused damage to his pituitary gland; this was responsible not only for his gain in weight and loss of body hair but also for undermining his hormonally driven aggressiveness and assertiveness, and hence for his abnormal submissiveness and placidity.

The diencephalon is especially a regulator of basic functions—of sleep, of appetite, of libido. And all of these were at a low ebb with Greg—he had (or expressed) no sexual interest; he did not think of eating, or express any desire to eat, unless food was brought to him. He seemed to exist only in the present, only in response to the immediacy of stimuli around him. If he was not stimulated, he fell into a sort of daze.

Left alone, Greg would spend hours in the ward without spontaneous activity. This inert state was at first described by the nurses as "brooding"; it had been seen in the temple as "meditating"; my own feeling was that it was a profoundly pathological mental "idling," almost devoid of mental content or affect. It was difficult to give a name to this state, so different from alert, attentive wakefulness, but also, clearly, quite different from sleep—it had a blankness resembling no normal state. It reminded me somewhat of the vacant states I had seen with some of my postencephalitic patients and, as with them, went with profound damage to the diencephalon. As soon as one talked to him, or if he was stimulated by sounds (especially music) near him, he would "come to," "awaken," in an astonishing way.

Once Greg was "awakened," once his cortex came to life, one saw that his animation itself had a strange quality—an uninhibited and quirky quality of the sort one tends to see when the orbital portions of the frontal lobes (that is, the portions adjacent to the eyes) are damaged, a so-called orbitofrontal syndrome. The frontal lobes are the most complex part of the brain, concerned not with the "lower" functions of movement and sensation, but the highest ones of integrating all judgment and behavior, all imagination and emotion, into that unique identity that we like to speak of as "personality" or "self." Damage to other parts of the brain may produce specific disturbances of sensation or movement, of language, or of specific perceptual, cognitive, or memory functions. Damage to the frontal lobes, in contrast, does not affect these, but produces a subtler and profounder disturbance of identity.

And it was this—rather than his blindness, or his weakness, or his disorientation, or his amnesia—that so horrified his parents when they finally saw Greg in 1975. It was not just that he was damaged, but that he was changed beyond recognition, had been "dispossessed," in his father's words, by a sort of simulacrum, or changeling, which had Greg's voice and manner and humor and intelligence but not his "spirit" or "realness" or "depth"—a changeling whose wisecracking and levity formed a shocking counterpoint to the fearful gravity of what had happened.

This sort of wisecracking, indeed, is quite characteristic of such orbito-frontal syndromes—and is so striking that it has been given a name unto itself: *witzelsucht*, or "joking disease." Some restraint, some caution, some inhibition, is destroyed, and patients with such syndromes tend to react immediately and incontinently to everything around them and everything within them—to virtually every object, every person, every sensation, every word, every thought, every emotion, every nuance and tone.

There is an overwhelming tendency, in such states, to wordplay and puns. Once when I was in Greg's room another patient walked past. "That's Bernie," I said. "Bernie the Hernie," quipped Greg. Another day when I visited him, he was in the dining room, awaiting lunch. When a nurse announced, "Lunch is here," he immediately responded, "It's time for cheer"; when she said, "Shall I take the skin off your chicken?" he instantly responded, "Yeah, why don't you slip me some skin." "Oh, you want the skin?" she asked, puzzled.

"Nah," he replied, "it's just a saying." He was, in a sense, preternaturally sensitive—but it was a sensitivity that was passive, without selectivity or focus. There is no differentiation in such a sensitivity—the grand, the trivial, the sublime, the ridiculous, are all mixed up and treated as equal.[5] There may be a childlike spontaneity and transparency about such patients in their immediate and unpremeditated (and often playful) reactions. And yet there is something ultimately disquieting, and bizarre, because the reacting mind (which may still be highly intelligent and inventive) loses its coherence, its inwardness, its autonomy, its "self," and becomes the slave of every passing sensation. The French neurologist François Lhermitte speaks of an "environmental dependency syndrome" in such patients, a lack of psychological distance between them and their environment. So it was with Greg: he seized his environment, he was seized by it, he could not distinguish himself from it.[6]

Dreaming and waking, for us, are usually distinct—dreaming is enclosed in sleep and enjoys a special license because it is cut off from external perception and action; while waking perception is constrained by reality.[7] But in

[5] Luria provides immensely detailed, at times almost novelistic, descriptions of frontal lobe syndromes—in *Human Brain and Psychological Processes*—and sees this "equalization" as the heart of such syndromes.

[6] A similar indiscriminate reactivity is sometimes seen in people with Tourette's syndrome—sometimes in the automatic form of echoing others' words or actions, sometimes in the more complex forms of mimicry, parodying or impersonating others' behavior, or in incontinent verbal associations (rhymings, punnings, clangings).

[7] Rodolfo Llinás and his colleagues at New York University, comparing the electrophysiological properties of the brain in waking and dreaming, postulate a single fundamental mechanism for both—a ceaseless inner talking between cerebral cortex and thalamus, a ceaseless interplay of image and feeling, irrespective of whether there is sensory input or not. When there is sensory input, this interplay integrates it to generate waking consciousness, but in the absence of sensory input it continues to generate brain states, those brain states we call fantasy, hallucination, or dreams. Thus waking consciousness is dreaming—but dreaming constrained by external reality.

Greg the boundary between waking and sleep seemed to break down, and what emerged was a sort of waking or public dream, in which dreamlike fancies and associations and symbols would proliferate and weave themselves into the waking perceptions of the mind.[8] These associations were often startling and sometimes surrealistic in quality. They showed the power of fancy at play and, specifically, the mechanisms—displacement, condensation, "overdetermination," and so on—that Freud has shown to be characteristic of dreams.

One felt all this very strongly with Greg; that he was often in some intermediate, half-dreamlike state in which, if the normal control and selectivity of thinking was lost, there was a half freedom, half compulsion, of fantasy and wit. To see this as pathological was necessary but insufficient: it had elements of the primitive, the childlike, the playful. Greg's absurdist, often gnomic utterances, along with his seeming serenity (actually blandness), gave him an appearance of innocence and wisdom combined, gave him a special status on the ward, ambiguous but respected, a Holy Fool.

Though as a neurologist I had to speak of Greg's "syndrome," his "deficits," I did not feel this was adequate to describe him. I felt, one felt, that he had become another "kind" of person; that though his frontal lobe damage had taken away his identity in a way, it had also given him a sort of identity or personality, albeit of an odd and perhaps a primitive sort.

If Greg was alone, in a corridor, he seemed scarcely alive; but as soon as he was in company, he was a different person altogether. He would "come to," he would be funny, charming, ingenuous, sociable. Everyone liked him; he would re-

---

[8] Dreamlike or oneiric states have been described, by Luria and others, with lesions of the thalamus and diencephalon. J.-J. Moreau, in a famous early study, *Hashish and Mental Illness* (1845), described both madness and hashish trances as "waking dreams." A particularly striking form of waking dream may be seen with the severer forms of Tourette's syndrome, where the external and the internal, the perceptual and the instinctual, burst forth in a sort of public phantasmagoria or dream.

spond to anyone at once, with a lightness and a humor and an absence of guile or hesitation; and if there was something too light or flippant or indiscriminate in his interactions and reactions, and if, moreover, he lost all memory of them in a minute, well, there were worse things; it was understandable, one of the results of his disease. Thus one was very aware, in a hospital for chronic patients like ours, a hospital where feelings of melancholy, of rage, and of hopelessness simmer and preside, of the virtue of a patient such as Greg—who never appeared to have bad moods, and who, when activated by others, was invariably cheerful, euphoric.

He seemed, in an odd way, and in consequence of his sickness, to have a sort of vitality or health—a cheeriness, an inventiveness, a directness, an exuberance, which other patients, and indeed the rest of us, found delightful in small doses. And where he had been so "difficult," so tormented, so rebellious in his pre-Krishna days, all this anger and torment and angst now seemed to have vanished; he seemed to be at peace. His father, who had had a terrible time in Greg's stormy days, before he got "tamed" by drugs, by religion, by tumor, said to me in an unbuttoned moment, "It's like he had a lobotomy," and then, with great irony, "Frontal lobes—who needs 'em?"

One of the most striking peculiarities of the human brain is the great development of the frontal lobes—they are much less developed in other primates and hardly evident at all in other mammals. They are the part of the brain that grows and develops most after birth (and their development is not complete until about the age of seven). But our ideas about the function of the frontal lobes, and the role they play, have had a tortuous and ambiguous history and are still far from clear. These uncertainties are well exemplified by the famous case of Phineas Gage, and the interpretations and misinterpretations, from 1848 to the present, of his case. Gage was the very capable foreman of a gang of workers constructing a railroad line near Burlington, Vermont, when a bizarre accident befell him in September 1848. He was setting an explosive charge, using a tamping iron (a crowbarlike instrument weighing thirteen pounds and more than a yard long), when the charge went off prematurely, blowing the tamping iron straight through his head. Though he was knocked down, incredibly he was not killed but only stunned for a moment. He was able to get up and take a cart into town. There he appeared perfectly rational and calm and alert and greeted the local doctor by saying, "Doctor, here is business enough for you."

Soon after his injury, Gage developed a frontal lobe abscess and fever, but this resolved within a few weeks, and by the beginning of 1849 he was called "completely recovered." That he had survived at all was seen as a medical miracle, and that he was seemingly unchanged after sustaining huge damage to the frontal lobes of the brain seemed to support the idea that these were either functionless or had no functions that could not be performed equally by the remaining, undamaged portions of the brain. Where phrenologists, earlier in the century, had seen every part of the brain surface as the "seat" of a particular intellectual or moral faculty, a reaction to this had set in during the 1830s and 1840s, to such an extent that the brain was sometimes seen as being as undifferentiated as the liver. Indeed, the great physiologist Flourens had said, "The brain secretes thought as the liver secretes bile." The apparent absence of any change in Gage's behavior seemed to support this notion.

Such was the influence of this doctrine that, despite clear evidence from other sources of a radical change in Gage's "character" within weeks of the accident, it was only twenty years later that the physician who had studied him most closely, John Martyn Harlow (now, apparently, moved by the new doctrines of "higher" and "lower" levels in the nervous system, the higher inhibiting or constraining the lower) provided a vivid description of all that he had ignored, or at least not mentioned, in 1848:

[Gage is] fitful, irreverent, indulging at times in the grossest profanity (which was not previously his custom), manifesting but little deference for his fellows, impatient of restraint or advice when it conflicts with his desires, at times pertinaciously obstinate, yet capricious and vacillating, devising many plans of future operations, which are no sooner arranged than they are abandoned in turn for others appearing more feasible. A child in his intellectual capacity and manifestations, he has the animal passions of a strong man. Previous to his injury, although untrained in the schools, he possessed a well-balanced mind, and was looked upon by those who knew him as a shrewd, smart businessman, very energetic and persistent in executing all his plans of operation. In this regard his mind was radically changed, so decidedly that his friends and acquaintances said he was "no longer Gage."

It seemed that a sort of "disinhibition" had occurred with the frontal lobe injury, releasing something animal-like or childlike, so that Gage now became a slave of his immediate whims and impulses, of what was immediately around him, without the deliberation, the consideration of past and future, that had marked him in the past, or his previous concern for others and the consequences of his actions.[9]

But excitement, release, disinhibition, are not the only possible effects of frontal lobe damage. David Ferrier (whose Gulstonian Lectures of 1879 introduced the Gage case to a worldwide medical community) observed a different sort of syndrome in 1876, when he removed the frontal lobes of monkeys:

Notwithstanding this apparent absence of physiological symptoms, I could perceive a very decided alteration in

---

[9] Robert Louis Stevenson wrote *The Strange Case of Dr. Jekyll and Mr. Hyde* in 1886. It is not known whether he knew of the Gage case, though this had become common knowledge since the early 1880s—but he was assuredly moved by the Jacksonian doctrine of higher and lower levels in the brain, the notion that it was only our "higher" (and perhaps fragile) intellectual centers that held back the animal propensities of the "lower."

the animal's character and behaviour. . . . Instead of, as before, being actively interested in their surroundings, and curiously prying into all that came within the field of their observation, they remained apathetic, or dull, or dozed off to sleep, responding only to the sensations or impressions of the moment, or varying their listlessness with restless and purposeless wanderings to and fro. While not actually deprived of intelligence, they had lost, to all appearance, the faculty of attentive and intelligent observation.

In the 1880s it became apparent that tumors of the frontal lobes could produce symptoms of many sorts: sometimes listlessness, hebetude, slowness of mental activity, sometimes a definite change in character and loss of self-control—sometimes even (according to Gowers) "chronic insanity." The first operation for a frontal lobe tumor was performed in 1884, and the first frontal lobe operation for purely psychiatric symptoms was done in 1888. The rationale here was that in these (probably schizophrenic) patients, the obsessions, the hallucinations, the delusional excitements, were due to overactivity, or pathological activity, in the frontal lobes.

There was to be no repetition of such forays for forty-five years, until the 1930s, when the Portuguese neurologist Egas Moniz devised the operation he called "prefrontal leucotomy" and immediately applied this to twenty patients, some with anxiety and depression, some with chronic schizophrenia. The results he claimed aroused huge interest when his monograph was published in 1936, and his lack of rigor, his recklessness, and perhaps dishonesty were all overlooked in the flush of therapeutic enthusiasm. Moniz's work led to an explosion of "psychosurgery" (the term he had coined) all over the world—Brazil, Cuba, Romania, Great Britain, and especially Italy—but its greatest resonance was to be in the United States, where the neurologist Walter Freeman invented a horrible new form of surgical approach that he called transorbital lobotomy. He described the procedure as follows:

This consists of knocking them out with a shock and while they are under the "anesthetic" thrusting an ice pick up between the eyeball and the eyelid through the roof of the orbit actually into the frontal lobe of the brain and making the lateral cut by swinging the thing from side to side. I have done two patients on both sides and another on one side without running into any complications, except a very black eye in one case. There may be trouble later on but it seemed fairly easy, although definitely a disagreeable thing to watch. It remains to be seen how these cases hold up, but so far they have shown considerable relief of their symptoms, and only some of the minor behavior difficulties that follow lobotomy. They can even get up and go home within an hour or so.

The ease of doing psychosurgery as an office procedure, with an ice pick, aroused not consternation and horror, as it should have, but emulation. More than ten thousand operations had been done in the United States by 1949, and a further ten thousand in the two years that followed. Moniz was widely acclaimed as a "savior" and received the Nobel Prize in 1951—the climax, in Macdonald Critchley's words, of "this chronicle of shame."

What was achieved, of course, was never "cure," but a docile state, a state of passivity, as far (or farther) from "health" than the original active symptoms, and (unlike these) with no possibility of resolution or reversal. Robert Lowell, in "Memories of West Street and Lepke," writes of the lobotomized Lepke:

> Flabby, bald, lobotomized,
> he drifted in a sheepish calm,
> where no agonizing reappraisal
> jarred his concentration on the electric chair—
> hanging like an oasis in his air
> of lost connections. . . .

When I worked at a state psychiatric hospital between 1966 and 1990, I saw dozens of these pathetic lobotomized patients, many far more damaged even than Lepke, some psychically dead, murdered, by their "cure."[10]

Whether or not there are in the frontal lobes a mass of pathological circuits causing the torments of mental illness—the simplistic notion first put forward in the 1880s, and embraced by Moniz—there is certainly a downside to their great and positive powers. The weight of consciousness and conscience and conscientiousness itself, the weight of duty, obligation, responsibility, can press on us sometimes with unbearable force, so that we long for a release from its crushing inhibitions, from sanity and sobriety. We long for a holiday from our frontal lobes, a Dionysiac fiesta of sense and impulse. That this is a need of our constrained, civilized, hyperfrontal nature has been recognized in every time and culture. All of us need to take little holidays from our frontal lobes—the tragedy is when, through grave illness or injury, there is no return from the holiday, as with Phineas Gage, or with Greg.[11]

---

[10] The huge scandal of leucotomy and lobotomy came to an end in the early fifties, not because of any medical reservation or revulsion, but because a new tool—tranquillizers—had now become available, which purported (as had psychosurgery itself) to be wholly therapeutic and without adverse effects. Whether there is that much difference, neurologically or ethically, between psychosurgery and tranquillizers is an uncomfortable question that has never been really faced. Certainly the tranquillizers, if given in massive doses, may, like surgery, induce "tranquillity," may still the hallucinations and delusions of the psychotic, but the stillness they induce may be like the stillness of death—and, by a cruel paradox, deprive patients of the natural resolution that may sometimes occur with psychoses and instead immure them in a lifelong, drug-caused illness.

[11] Though the medical literature of frontal lobe syndromes starts with the case of Phineas Gage, there are earlier descriptions of altered mental states not identifiable at the time—which we can now, in retrospect, see as frontal lobe syndromes. One such account is related by Lytton Strachey in "The Life, Illness, and Death of Dr. North." Dr. North, a master of Trinity College, Cambridge, in the eighteenth century, was a man with severe anxieties and tormenting obsessional traits, who was hated and dreaded by the fellows of the college for his punctiliousness, his moralizing, and his merciless severity. Until one day, in college, he suffered a stroke:

In a March 1979 note about Greg, I reported that "games, songs, verses, converse, etc. hold him together completely . . . because they have an organic rhythm and stream, a flowing of being, which carries and holds him." I was strongly reminded here of what I had seen with my amnesiac patient Jimmie, how he seemed held together when he attended Mass, by his relationship to and participation in an act of meaning, an organic unity, which overrode or bypassed the disconnections of his amnesia.[12] And what I had observed with a patient in England, a musicologist with profound amnesia from a temporal lobe encephalitis, unable to remember events or facts for more than a few seconds, but able to remember, and indeed to learn, elaborate musical pieces, to conduct them, to perform them, and even to improvise at the organ.[13]

It was similar with Greg as well: he not only had an excellent memory for songs of the sixties, but was able to learn

---

His recovery was not complete; his body was paralyzed on the left side; but it was in his mind that the most remarkable change occurred. His fears had left him. His scrupulosity, his diffidence, his seriousness, even his morality—all had vanished. He lay on his bed, in reckless levity, pouring forth a stream of flippant observations, and naughty stories, and improper jokes. While his friends hardly knew which way to look, he laughed consumedly, his paralyzed features drawn up in a curiously distorted grin. . . . Attacked by epileptic seizures, he declared that the only mitigation of his sufferings lay in the continued consumption of wine. He, who had been so noted for his austerity, now tossed off, with wild exhilaration, glass after glass of the strongest sherry.

Strachey gives us here a precise and beautifully described picture of a frontal lobe stroke altering the personality in a major and, so to speak, "therapeutic" way.

[12] The nature of the "organic unity," at once dynamic and semantic, which is central to music, incantation, recitation, and all metrical structures, has been most profoundly analyzed by Victor Zuckerkandl in his remarkable book *Sound and Symbol*. It is typical of such flowing dynamic-semantic structures that each part leads on to the next, that every part has reference to the rest. Such structures cannot usually be perceived, or remembered, in part—they are perceived and remembered, if at all, as wholes.

[13] This patient is the subject of a remarkable BBC film made by Jonathan Miller, *Prisoner of Consciousness* (November 1988).

new songs easily, despite his difficulty in retaining any "facts." It seemed as if wholly different kinds—and mechanisms—of memory might be involved. Greg was also able to pick up limericks and jingles with ease (and had indeed picked up hundreds of these from the radio and television that were always on in the ward). Soon after his admission, I tested him with the following limerick:

> Hush-a-bye baby,
> Hush quite a lot,
> Bad babies get rabies
> And have to be shot.

Greg immediately repeated this, without error, laughed at it, asked if I'd made it up, and compared it with "something gruesome, like Edgar Allan Poe." But two minutes later he could not recall it, until I reminded him of the underlying rhythm. With a few more repetitions, he learned it without cueing and thereafter recited it whenever he met me.

Was this facility for learning jingles and songs a mere procedural or performative one, or could it provide emotional depth or generalizability of a sort that Greg did not normally have access to? There seemed no doubt that some music could move him profoundly, could be a door to depths of feeling and meaning to which he normally had no access, and one felt Greg was a different person at these times. He no longer seemed to have a frontal lobe syndrome, but was (so to speak) temporarily "cured" by the music. Even his EEG, so slow and incoherent most of the time, became calm and rhythmical with music.[14]

---

[14] Another patient in Williamsbridge, Harry S.—a gifted man, a former engineer—suffered a huge cerebral hemorrhage from a burst aneurysm, with gross destruction of both frontal lobes. Emerging from a coma, he started to recover and eventually recovered most of his former intellectual powers, but remains, like Greg, severely impaired—bland, flat, indifferent emotionally. But all this changes, suddenly, when he sings. He has a fine tenor voice and loves Irish songs. When he sings, he does so with a fullness of feeling, a tenderness, a lyricism, that are astounding—the more so because one sees no hint of this at any

It is easy to show that simple information can be embedded in songs; thus we can give Greg the date every day in the form of a jingle, and he can readily isolate this and say it when asked, without the jingle. But what does it mean to say, "This is July 9, 1995," when one is sunk in the profoundest amnesia, when one has lost a sense of time and history, when one is existing from moment to moment in a sequenceless limbo? Knowing the date means nothing in these circumstances. Could one, however, through the evocativeness and power of music, perhaps using songs with specially written lyrics—songs that relate something valuable about himself or the current world—accomplish something more lasting, deeper? Give Greg not only the "facts," but a sense of time and history, of the relatedness of events, an entire (if artificial) framework for thinking and feeling?

It seemed natural, at this time, given Greg's blindness and the revelation of his potential for learning, that he should be given an opportunity to learn Braille. Arrangements were made with the Jewish Institute for the Blind for him to enter intensive training, four times a week. It should not have been a disappointment, nor indeed a surprise, that Greg was unwilling to learn any Braille—that he was startled and bewildered at finding this imposed on him, and cried out, "What's going on? Do you think I'm blind? Why am I here, with blind people all around me?" Attempts were made to explain things to him, and he responded, with impeccable logic, "If I were blind, I would be the first person to know it." The institute said they had never had such a difficult patient, and the project was quietly allowed to drop. And indeed, with the failure of the Braille program, a sort of hopelessness gripped us, and perhaps Greg, too. We could do nothing, we felt; he had no potential for change.

---

other time and might well think his emotional capacity entirely destroyed. He shows every emotion appropriate to what he sings—the frivolous, the jovial, the tragic, the sublime—and seems to be transformed while he sings.

Greg by this time had had several psychological and neuro-psychological evaluations, and these, besides commenting on his memory and attentional problems, had all spoken of him as being "shallow," "infantile," "insightless," "euphoric." It was easy to see why these words had been used; Greg was like this for much of the time. But was there a deeper Greg beneath his illness, beneath the shallowing effect of his frontal lobe loss and amnesia? Early in 1979, when I questioned him, he said he was "miserable . . . at least in the corporeal part," and added, "It's not much of a life." At such times, it was clear that he was not just frivolous and euphoric, but capable of deep, and indeed melancholic, reactions to his plight. The comatose Karen Ann Quinlan was then very much in the news, and each time her name and fate were mentioned, Greg became distressed and silent. He could never tell me, explicitly, why this so interested him—but it had to be, I felt, because of some sort of identification of her tragedy with his own. Or was this just his incontinent sympathy, his falling at once into the mood of any stimulus or news, falling almost helplessly, mimetically, into its mood?

This was not a question I could decide at first, and perhaps, too, I was prejudiced against finding any depths in Greg, because the neuropsychological studies I knew of seemed to disallow this possibility. But these studies were based on brief evaluations, not on long-continued observation and relationship of a sort that is, perhaps, only possible in a hospital for chronic patients, or in situations where a whole world, a whole life, are shared with the patient.

Greg's "frontal lobe" characteristics—his lightness, his quick-fire associations—were fun, but beyond this there shone through a basic decency and sensitivity and kindness. One felt that Greg, though damaged, still had a personality, an identity, a soul.[15]

---

[15] Mr. Thompson ("A Matter of Identity"), who also had both amnesia and a frontal lobe syndrome, by contrast often seemed "desouled." In him the wise-cracking was manic, ferocious, frenetic, and relentless; it rushed on like a torrent,

When he came to Williamsbridge we all responded to his intelligence, his high spirits, his wit. All sorts of therapeutic programs and enterprises were started at this time, but all of them—like the learning of Braille—ended in failure. The sense of Greg's incorrigibility gradually grew on us, and with this we started to do less, to hope less. Increasingly, he was left to his own devices. He slowly ceased to be a center of attention, the focus of eager therapeutic activities—more and more he was left to himself, left out of programs, not taken anywhere, quietly ignored.

It is easy, even if one is not an amnesiac, to lose touch with current reality in the back wards of hospitals for the chronically ill. There is a simple round that has not changed in twenty, or fifty, years. One is wakened, fed, taken to the toilet, and left to sit in a hallway; one has lunch, one is taken to bingo, one has dinner and goes to bed. The television may indeed be left on, blaring, in the television room—but most patients pay no attention to it. Greg, it is true, enjoyed his favorite soap operas and westerns and learned an enormous number of advertising jingles by heart. But the news, for the most part, he found boring and, increasingly, unintelligible. Years can pass, in a sort of timeless limbo, with few, and certainly no memorable, markers of the passage of time.

As ten years or so went by, Greg showed a complete absence of development, his talk seemed increasingly dated and repertorial, for nothing new was being added to it, or him. The tragedy of his amnesia seemed to become greater with the years, although his amnesia itself, his neurological syndrome, remained much the same.

---

oblivious to tact, to decency, to propriety, to everything, including the feelings of everyone around him. Whether Greg's (at least partial) preservation of ego and identity was due to the lesser severity of his syndrome, or to underlying personality differences, is not wholly clear. Mr. Thompson's premorbid personality was that of a New York cabbie, and in some sense his frontal lobe syndrome merely intensified this. Greg's personality was gentler, more childlike, from the start—and this, it seemed to me, even colored his frontal lobe syndrome.

In 1988 Greg had a seizure—he had never had one before (although he had been on anticonvulsants, as a precaution, since the time of his surgery)—and in the seizure broke a leg. He did not complain of this, he did not even mention it; it was only discovered when he tried to stand up the following day. He had, apparently, forgotten it as soon as the pain eased and as soon as he had found a comfortable position. His not knowing that he had broken a leg seemed to me to have similarities to his not knowing he was blind, his inability, with his amnesia, to hold in mind an absence. When the leg caused pain, briefly, he knew something had happened, he knew it was there; as soon as the pain ceased, it went from his mind. Had he had visual hallucinations or phantoms (as the blind sometimes do, at least in the first months and years after losing their sight), he could have spoken of them, said, "Look!" or "Wow!" But in the absence of actual visual input, he could hold nothing in mind about seeing, or not-seeing, or the loss of a visual world. In his person, and in his world, now, Greg knew only presence, not absence. He seemed incapable of registering any loss—loss of function in himself, or of an object, or a person.

In June of 1990, Greg's father, who had come every morning before work to see Greg and would joke and chat with him for an hour, suddenly died. I was away at the time (mourning my own father), and hearing the news of Greg's bereavement on my return, I hastened to see him. He had been given the news, of course, when it happened. And yet I was not quite sure what to say—had he been able to absorb this new fact? "I guess you must be missing your father," I ventured.

"What do you mean?" Greg answered. "He comes every day. I see him every day."

"No," I said, "he's no longer coming. . . . He has not come for some time. He died last month."

Greg flinched, turned ashen, became silent. I had the impression he was shocked, doubly shocked, at the sudden, appalling news of his father's death, and at the fact that he himself did not know, had not registered, did not remember. "I guess he must have been around fifty," he said.

"No, Greg," I answered, "he was well up in his seventies."

Greg grew pale again as I said this. I left the room briefly; I felt he needed to be alone with all this. But when I returned a few minutes later, Greg had no memory of the conversation we had had, of the news I had given him, no idea that his father had died.

Very clearly, at least, Greg showed a capacity for love and grief. If I had ever doubted Greg's capacity for deeper feeling, I no longer doubted it now. He was clearly devastated by his father's death—he showed nothing "flip," no levity, at this time.[16] But would he have the ability to mourn? Mourning requires that one hold the sense of loss in one's mind, and it was far from clear to me that Greg could do this. One might indeed tell him that his father had died, again and again. And every time it would come as something shocking and new and cause immeasurable distress. But then, in a few minutes, he would forget and be cheerful again, and was so prevented from going through the work of grief, the mourning.[17]

I made a point of seeing Greg frequently in the following months, but I did not again bring up the subject of his father's death. It was not up to me, I thought, to confront him with this—indeed it would be pointless and cruel to do so; life itself, surely, would do so, for Greg would discover his father's absence.

I made the following note on November 26, 1990: "Greg shows no conscious knowing that his father has died—when asked where his father is, he may say, 'Oh, he went down to

---

[16] This is in distinction to Mr. Thompson, who with his more severe frontal lobe syndrome had been reduced to a sort of nonstop, wisecracking, talking machine, and when told of his brother's death quipped "He's always the joker!" and rushed on to other, irrelevant things.

[17] The amnesiac musicologist in the BBC film *Prisoner of Consciousness* showed something both similar and different. Every time his wife went out of the room, he had a sense of calamitous, permanent loss. When she came back, five minutes later, he sobbed with relief, saying, "I thought you were dead."

the patio,' or 'He couldn't make it today,' or something else plausible. But he no longer wants to go home, on weekends, on Thanksgiving, as he so loved to—he must find something sad or repugnant in the fatherless house now, even though he cannot (consciously) remember or articulate this. Clearly he has established an association of sadness."

Toward the end of the year Greg, normally a sound sleeper, started to sleep poorly, to get up in the middle of the night and wander gropingly for hours around his room. "I've lost something, I'm looking for something," he would say when asked— but what he had lost, what he was looking for, he could never explain. One could not avoid the feeling that Greg was looking for his father, even though he could give no account of what he was doing and had no explicit knowledge of what he had lost. But, it seemed to me, there was perhaps now an implicit knowledge and perhaps, too, a symbolic (though not a conceptual) knowing.

Greg had seemed so sad since his father's death that I felt he deserved a special celebration—and when I heard, in August of 1991, that his beloved group, the Grateful Dead, would be playing at Madison Square Garden in a few weeks, this seemed just the thing. Indeed, I had met one of the drummers in the band, Mickey Hart, earlier in the summer, when we had both testified before the Senate about the therapeutic powers of music, and he made it possible for us to obtain tickets at the last minute, to bring Greg, wheelchair and all, into the concert, where a special place would be saved for him near the soundboard, where acoustics were best.

We made these arrangements at the last minute, and I had given Greg no warning, not wanting to disappoint him if we failed to get seats. But when I picked him up at the hospital and told him where we were going, he showed great excitement. We got him dressed swiftly and bundled him into the car. As we got into midtown, I opened the car windows, and the sounds and smells of New York came in. As we cruised down Thirty-third Street, the smell of hot pretzels suddenly struck him; he inhaled deeply and laughed. "That's the most New York smell in the world."

There was an enormous crowd converging on Madison Square Garden, most in tie-dyed T-shirts—I had hardly seen a tie-dyed T-shirt in twenty years, and I myself began to think we were back in the sixties, or perhaps that we had never left them. I was sorry that Greg could not see this crowd; he would have felt himself one of them, at home. Stimulated by the atmosphere, Greg started to talk spontaneously—very unusual for him—and to reminisce about the sixties:

Yeah, there were the be-ins in Central Park. They haven't had one for a long time—over a year, maybe, can't remember exactly. . . . Concerts, music, acid, grass, everything. . . . First time I was there was Flower-Power Day. . . . Good times . . . lots of things started in the sixties—acid rock, the be-ins, the love-ins, smoking. . . . Don't see it much these days. . . . Allen Ginsberg—he's down in the Village a lot, or in Central Park. I haven't seen him for a long time. It's over a year since I last saw him. . . .

Greg's use of the present tense, or the near-present tense; his sense of all these events, not as far distant, much less as terminated, but as having taken place "a year ago, maybe" (and, by implication, likely to take place again, at any time); all this, which seemed so pathological, so anachronistic in clinical testing, seemed almost normal, natural, now that we were part of this sixties crowd sweeping toward the Garden.

Inside the Garden we found the special place reserved for Greg's wheelchair near the soundboard. And now Greg was growing more excited by the minute; the roar of the crowd excited him—"It's like a giant animal," he said—and the sweet, hash-laden air. "What a great smell," he said, inhaling deeply. "It's the least stupid smell in the world."[18]

---

[18] Jean Cocteau, in fact, said this of opium. Whether Greg was quoting this, consciously or unconsciously, I do not know. Smells are sometimes even more evocative than music; and the percepts of smells, generated in a very primitive part of

As the band came onstage, and the noise of the crowd grew greater, Greg was transported by the excitement and started clapping loudly and shouting in an enormous voice, "Bravo! Bravo!" then "Let's go!" followed by "Let's go, Hypo," followed, homophonously, by "Ro, Ro, Ro, Harry-Bo." Pausing a moment, Greg said to me, "See the tombstone behind the drums? See Jerry Garcia's Afro?" with such conviction that I was momentarily taken in and looked (in vain) for a tombstone behind the drums—before realizing it was one of Greg's confabulations—and at the now-grey hair of Jerry Garcia, which fell in a straight, unhindered descent to his shoulders.

And then, "Pigpen!" Greg exclaimed, "You see Pigpen there?"

"No," I replied, hesitantly, not knowing how to reply. "He's not there. . . . You see, he's not with the Dead anymore."

"Not with them?" said Greg, in astonishment. "What happened—he got busted or something?"

"No, Greg, not busted. He died."

"That's awful," Greg answered, shaking his head, shocked. And then a minute later, he nudged me again. "Pigpen! You see Pigpen there?" And, word for word, the whole conversation repeated itself.

But then the thumping, pounding excitement of the crowd got him—the rhythmic clapping and stamping and chanting possessed him—and he started to chant, "The Dead! The Dead!" then with a shift of rhythm, and a slow emphasis on each word, "We want the Dead!" And then, "Tobacco Road, Tobacco Road," the name of one of his favorite songs, until the music began.

The band began with an old song, "Iko, Iko," and Greg joined in with gusto, with abandon, clearly knowing all the

---

the brain—the "smell brain," or rhinencephalon—may not go through the complex, multistage memory systems of the medial temporal lobe. Olfactory memories, neurally, are almost indelible; thus they may be remembered despite an amnesia. It would be fascinating to bring Greg hot pretzels, or hash, to see whether their smells could evoke memories of the concert. He himself, the next day, spontaneously mentioned the "great" smell of pretzels—it was very vivid for him—and yet he could not locate the smell in place or time.

words, and especially luxuriating in the African-sounding chorus. The whole vast Garden now was in motion with the music, eighteen thousand people responding together, everyone transported, every nervous system synchronized, in unison.

The first half of the concert had many earlier pieces, songs from the sixties, and Greg knew them, loved them, joined in. His energy and joy were amazing to see; he clapped and sang nonstop, with none of the weakness and fatigue he generally showed. He showed a rare and wonderful continuity of attention, everything orienting him, holding him together. Looking at Greg transformed in this way, I could see no trace of his amnesia, his frontal lobe syndrome—he seemed at this moment completely normal, as if the music was infusing him with its own strength, its coherence, its spirit.

I had wondered whether we should leave at the break midway through the concert—he was, after all, a disabled, wheelchair-bound patient, who had not really been out on the town, at a rock concert, for more than twenty years. But he said, "No, I want to stay, I want it all"—an assertion, an autonomy, I rejoiced to see and had hardly ever seen in his compliant life at the hospital. So we stayed, and in the interval went backstage, where Greg had a large hot pretzel and then met Mickey Hart and exchanged a few words with him. He had looked a little tired and pale before, but now he was flushed, excited by the encounter, charged and eager to be back for more music.

But the second half of the concert was somewhat strange for Greg: more of the songs dated from the mid- or late seventies and had lyrics that were unknown to him, though they were familiar in style. He enjoyed these, clapping and singing along wordlessly, or making up words as he went. But then there were newer songs, radically different, like "Picasso Moon," with dark and deep harmonies and an electronic instrumentation such as would have been impossible, unimaginable, in the 1960s. Greg was intrigued, but deeply puzzled. "It's weird stuff," he said, "I never heard anything like it before." He listened intently, all his musical senses stirred, but with a slightly scared and bewildered look, as if seeing a new animal, a new plant, a new world, for the first time. "I guess it's some new, experimental stuff," he said, "something they never played before. Sounds futuristic . . . maybe it's the music of the future." The newer songs he heard went far beyond any development that he could have imagined, were so beyond (and in some ways so unlike) what he associated with the Dead, that it "blew his mind." It was, he could not doubt, "their" music he was hearing, but it gave him an almost unbearable sense of hearing the future—as late Beethoven would have struck a devotee if it had been played at a concert in 1800.

"That was fantastic," he said, as we filed out of the Garden. "I will always remember it. I had the time of my life." I played CDs of the Grateful Dead in the car on the way home, to hold as long as possible the mood and memory of the concert. I feared that if I stopped playing the Dead, or talking about them, for a single moment, all memory of the concert would go from his mind. Greg sang along enthusiastically all the way back, and when we parted at the hospital, he was still in an exuberant concert mood.

But the next morning when I came to the hospital early, I found Greg in the dining room, alone, facing the wall. I asked him about the Grateful Dead—what did he think of them? "Great group," he said, "I love them. I heard them in Central Park and at the Fillmore East."

"Yes," I said, "you told me. But have you seen them since? Didn't you just hear them at Madison Square Garden?"

"No," he said, "I've never been to the Garden."[19]

---

[19] Greg has no recollection of the concert, seemingly—but when I was sent a tape of it, he immediately recognized some of the "new" pieces, found them familiar, was able to sing them. "Where did you hear that?" I asked as we listened to "Picasso Moon."

He shrugged uncertainly. But there is no doubt that he has learned it, nonetheless. I have taken now to visiting him regularly, with tapes of our concert and of the latest Grateful Dead concerts. He seems to enjoy the visits and has learned many of the new songs. And now, whenever I arrive, and he hears my voice, he lights up, and greets me as a fellow Deadhead.

A Surgeon's
Life

Tourette's syndrome is seen in every race, every culture, every stratum of society; it can be recognized at a glance once one is attuned to it; and cases of barking and twitching, of grimacing, of strange gesturing, of involuntary cursing and blaspheming, were recorded by Aretaeus of Cappadocia almost two thousand years ago. Yet it was not clinically delineated until 1885, when Georges Gilles de la Tourette, a young French neurologist—a pupil of Charcot's and a friend of Freud's—put together these historical accounts with observations of some of his own patients. The syndrome as he described it was characterized, above all, by convulsive tics, by involuntary mimicry or repetition of others' words or actions (echolalia and echopraxia), and by the involuntary or compulsive utterances of curses and obscenities (coprolalia). Some individuals (despite their affliction) showed an odd insouciance or nonchalance; some a tendency to make strange, often witty, occasionally dreamlike associations; some extreme impulsiveness and provocativeness, a constant testing of physical and social boundaries; some a constant, restless reacting to the environment, a lunging at and sniffing of everything or a sudden flinging of objects; and yet others an extreme stereotypy and obsessiveness—no two patients were ever quite the same.

Any disease introduces a doubleness into life—an "it," with its own needs, demands, limitations. With Tourette's, the "it" takes the form of explicit compulsion, a multitude of explicit impulsions and compulsions: one is driven to do this, to do that, against one's own will, or in deference to the alien will of the "it." There may be a conflict, a compromise, a collusion between these wills. Thus being "possessed" can be more than a figure of speech for someone with an impulse disorder like Tourette's, and no doubt in the Middle Ages it was sometimes literally seen as "possession." (Tourette himself was fascinated by the phenomenon of possession and wrote a play about the epidemic of demonic possession in medieval Loudun.)

But the relation of disease and self, "it" and "I," can be particularly complex in Tourette's, especially if it has been present from early childhood, growing up with the self, intertwining itself in every possible way. The Tourette's and the self shape themselves each to the other, come more and more to complement each other, until finally, like a long-married couple, they become a single, compound being. This relation is often destructive, but it can also be constructive, can add speed and spontaneity and a capacity for unusual and sometimes startling performance. For all its intrusiveness, Tourette's may be used creatively, too.

Yet in the years after its delineation, Tourette's tended to be seen not as an organic but as a "moral" disease—an expression of mischievousness or weakness of the will, to be treated by rectifying the will. From the 1920s to the 1960s, it tended to be seen as a psychiatric disease, to be treated by psychoanalysis or psychotherapy; but this, on the whole, proved ineffective, too. Then, with the demonstration, in the early 1960s, that the drug haloperidol could dramatically suppress its symptoms, Tourette's was regarded (in a sudden reversal) as a chemical disease, the result of an imbalance of a neurotransmitter, dopamine, in the brain. But all these views are partial, and reductive, and fail to do justice to the full complexity of Tourette's. Neither a biological nor a psychological nor a moral-social viewpoint is adequate; we must see Tourette's not only simultaneously from all three perspectives, but from an inner perspective, an existential perspective, that of the affected person himself. Inner and outer narratives here, as everywhere, must fuse.

Many professions, one would think, would be closed to someone with elaborate tics and compulsions or strange, antic behaviors, but this does not seem to be the case. Tourette's affects perhaps one person in a thousand, and we find people with Tourette's—sometimes the most severe Tourette's—in virtually every walk of life. There are Tourettic writers, mathematicians, musicians, actors, disc jockeys, construction workers, social workers, mechanics, athletes. Some things, one might think, would be completely out of the question— above all, perhaps, the intricate, precise, and steady work of a surgeon. This would have been my own belief not so long ago. But now, improbably, I know *five* surgeons with Tourette's.[1]

I first met Dr. Carl Bennett at a scientific conference on Tourette's in Boston. His appearance was unexceptionable—he was fiftyish, of middle size, with a brownish beard and mustache containing a hint of gray, and was dressed soberly in a dark suit—until he suddenly lunged or reached for the ground or jumped or jerked. I was struck both by his bizarre tics and by his dignity and calm. When I expressed incredulity about his choice of profession, he invited me to visit and stay with him, where he lived and practiced, in the town of Branford, in British Columbia—to do rounds at the hospital with him, to scrub with him, to see him in action. Now, four months later, in early October, I found myself in a small plane approaching Branford, full of curiosity and mixed expectations. Dr. Bennett met me at the airport, greeted me—a strange greeting, half lunge, half tic, a gesture of welcome idiosyncratically Touret-

---

[1] A further four surfaced (one an ophthalmic surgeon) following the original publication of this piece. In addition to these Tourettic surgeons, I now know of three Tourettic internists, two Tourettic neurologists, but only one Tourettic psychiatrist.

tized—grabbed my case, and led the way to his car in an odd, rapid skipping walk, with a skip each fifth step and sudden reachings to the ground as if to pick something up.

The situation of Branford is almost idyllic, nestled as it is in the shadow of the Rockies, in southeast British Columbia, with Banff and its mountains to the north, and Montana and Idaho to the south; it lies in a region of great gentleness and fertility but is ringed with mountains, glaciers, lakes. Bennett himself has a passion for geography and geology; a few years ago he took a year off from his surgical practice to study both at the University of Victoria. As he drove, he pointed out moraines, stratifications, and other formations, so that what had at first seemed to my eyes a mere pastoral landscape became charged with a sense of history and chthonic forces, of immense geological vistas. Such keen, fierce attention to every detail, such constant looking below the surface, such examination and analysis, are characteristic of the restless, questioning Tourettic mind. It is, so to speak, the other side of its obsessive and perseverative tendencies, its disposition to reiterate, to touch again and again.

And, indeed, whenever the stream of attention and interest was interrupted, Bennett's tics and iterations immediately reasserted themselves—in particular, obsessive touchings of his mustache and glasses. His mustache had constantly to be smoothed and checked for symmetry, his glasses had to be "balanced"—up and down, side to side, diagonally, in and out—with sudden, ticcy touchings of the fingers, until these, too, were exactly "centered." There were also occasional reachings and lungings with his right arm; sudden, compulsive touchings of the windshield with both forefingers ("The touching has to be symmetrical," he commented); sudden repositionings of his knees, or the steering wheel ("I have to have the knees symmetrical in relation to the steering wheel. They have to be *exactly* centered"); and sudden, high-pitched vocalizations, in a voice completely unlike his own, that sounded like "Hi, Patty," "Hi, there," and, on a couple of occasions, "Hideous!" (Patty, I learned later, was a former girlfriend, her name now enshrined in a tic.)[2]

There was little hint of this repertoire until we reached town and got obstructed by traffic lights. The lights did not annoy Bennett—we were in no hurry—but they did break up the driving, the kinetic melody, the swift, smooth stream of action, with its power to integrate mind and brain. The transition was very sudden: one minute, all was smoothness and action; the next, all was broken-upness, pandemonium, riot. When Bennett was driving smoothly, one had the feeling not that the Tourette's was in any way being suppressed but that the brain and the mind were in a quite different mode of action.

Another few minutes, and we had arrived at his house, a charming, idiosyncratic house with a wild garden, perched on a hill overlooking the town. Bennett's dogs, rather wolflike, with strange, pale eyes, barked, wagged their tails, bounded up to us as we drove in. As we got out of the car, he said "Hi, puppies!" in the same quick, odd, high, crushed voice he had earlier used for "Hi, Patty!" He patted their heads, a ticlike,

---

[2] Tics can have an ambiguous status, partway between meaningless jerks or noises and meaningful acts. Though the tendency to tic is innate in Tourette's, the particular *form* of tics often has a personal or historical origin. Thus a name, a sound, a visual image, a gesture, perhaps seen years before and forgotten, may first be unconsciously echoed or imitated and then preserved in the stereotypic form of a tic. Such tics are like hieroglyphic, petrified residues of the past and may indeed, with the passage of time, become so hieroglyphic, so abbreviated, as to become unintelligible (as "God be with you" was condensed, collapsed, after centuries, to the phonetically similar but meaningless "goodbye"). One such patient, whom I saw long ago, kept making an explosive, guttural, trisyllabic noise, which revealed itself, on analysis, as a very accelerated, crushed rendering of "*Verboten!*" in a convulsive parody of his father's constantly forbidding German voice.

A recent correspondent, a woman with Tourette's, after reading an earlier version of this piece, wrote that " 'enshrinement' . . . is the perfect word to describe the interplay between life and tics—the process by which the former gets incorporated into the latter. . . . It is almost as if the Tourettic body becomes an expressive archive—albeit jumbled—of one's life experience."

convulsive patting, a quick-fire volley of five pats to each, delivered with a meticulous symmetry and synchrony. "They're grand dogs, half-Eskimo, half-malamute," he said. "I felt I should get two of them, so they could companion each other. They play together, sleep together, hunt together—everything." And, I thought, are patted together: Did he get two dogs partly because of his own symmetrical, symmetrizing compulsions? Now, hearing the dogs bark, his sons ran out— two handsome teenage kids. I had a sudden feeling that Bennett might cry "Hi, kiddies!" in his Touretty voice and pat their heads, too, in synchrony, symmetrically. But he introduced them, Mark and David, individually to me. And then, as we entered the house, he introduced me to his wife, Helen, who was preparing a late-afternoon tea for all of us.

As we sat at the table, Bennett was repeatedly distracted by tics—a compulsive touching of the glass lampshade above his head. He had to tap the glass gently with the nails of both forefingers, to produce a sharp, half-musical click or, on occasion, a short salvo of clicks. A third of his time was taken up with this ticcing and clicking, which he seemed unable to stop. Did he have to do it? Did he have to sit there?

"If it were out of reach, would you still have to click it?" I asked.

"No," he said. "It depends entirely on how I'm situated. It's all a question of space. Where I am now, for example, I have no impulse to reach over to that brick wall, but if I were in range I'd have to touch it perhaps a hundred times." I followed his glance to the wall and saw that it was pockmarked, like the moon, from his touchings and jabbings; and, beyond it, the refrigerator door, dented and battered, as if from the impact of meteorites or projectiles. "Yeah," Bennett said, now following my glance. "I fling things—the iron, the rolling pin, the saucepan, whatever—I fling things at it if I suddenly get enraged." I digested this information in silence. It added a new dimension—a disquieting, violent one—to the picture I was building and seemed completely at odds with the genial, tranquil man before me.[3]

"If the light so disturbs you, why do you sit near it?" I asked.

"Sure, it's 'disturbance,' " Bennett answered. "But it's also stimulation. I like the feel and the sound of the click. But, yeah, it can be a great distraction. I can't study here, in the dining room—I have to go to my study, out of reach of the lamp."

The sense of personal space, of the self in relation to other objects and other people, tends to be markedly altered in Tourette's syndrome. I know many people with Tourette's who cannot tolerate sitting in a restaurant within touching distance of other people and may feel compelled, if they cannot avoid this, to reach out or lunge convulsively toward them. This intolerance may be especially great if the "provoking" person is behind the Touretter. Many people with Tourette's, therefore, prefer corners, where they are at a "safe" distance from others, and there is nobody behind them.[4] Analogous problems may arise, on occasion, when driving; there may be a sense that other vehicles are "too close" or "looming," even that they are suddenly "zooming," when they are (a non-Tourettic person would judge) at a normal distance. There

---

[3] Some people with Tourette's have flinging tics—sudden, seemingly motiveless urges or compulsions to throw objects—quite different from Bennett's flinging in rage. There may be a very brief premonition—enough, in one case, to yell a warning "Duck!"—before a dinner plate, a bottle of wine, or whatever is flung convulsively across the room. Identical throwing tics occurred in some of my postencephalitic patients when they were overstimulated by L-DOPA. (I see somewhat similar flinging behaviors—though not tics—in my two-year-old godson, now in a stage of primal antinomianism and anarchy.)

[4] This was comically shown on one occasion when I went to a restaurant for dinner with three Tourettic friends in Los Angeles. All three of them at once rushed for the corner seat—not, I think, in any competitive spirit, but because each saw it as an existential-neural necessity. The lucky one was able to sit calmly in his place, while the other two were constantly lunging at other diners behind them.

may also be, paradoxically, a tendency to be "attracted" to other vehicles, to drift or veer toward them—though the consciousness of this, and a greater speed of reaction, usually serves to avert any mishaps. (Similar illusions and urges, stemming from abnormalities in the neural basis of personal space, may occasionally be seen in parkinsonism, too.)

Another expression of Bennett's Tourette's—very different from the sudden impulsive or compulsive touching—is a slow, almost sensuous pressing of the foot to mark out a circle in the ground all around him. "It seems to me almost instinctual," he said when I asked him about it. "Like a dog marking its territory. I feel it in my bones. I think it is something primal, prehuman—maybe something that all of us, without knowing it, have in us. But Tourette's 'releases' these primitive behaviors."[5]

Bennett sometimes calls Tourette's "a disease of disinhibition." He says there are thoughts, not unusual in themselves, that anyone might have in passing but that are normally inhibited. With him, such thoughts perseverate in the back of the mind, obsessively, and burst out suddenly, without his consent or intention. Thus, he says, when the weather is nice he may want to be out in the sun getting a tan. This thought will be in the back of his mind while he is seeing his patients in the hospital and will emerge in sudden, involuntary utterances. "The nurse may say, 'Mr. Jones has abdominal pain,' and I'm looking out of the window saying, 'Tanning rays, tanning rays.' It might come out five hundred times in a morning. People in the ward must hear it—they can't *not* hear it—but I guess they ignore it or feel that it doesn't matter."

Sometimes the Tourette's manifests itself in obsessive

---

[5] Tourette's should not be regarded as a psychiatric disorder, but as a neurobiological disorder of a hyperphysiological sort, in which there may occur subcortical excitation and spontaneous stimulation of many phylogenetically primitive centers in the brain. A similar stimulation or release of "primitive" behaviors may be seen with the excitatory lesions of encephalitis lethargica, such as I describe in *Awakenings* (pp. 55–6). These were often apparent in the early days of the illness and became prominent again with the stimulation of L-DOPA.

thoughts and anxieties. "If I'm worried about something," Bennett told me as we sat around the table, "say, I hear a story about a kid being hurt, I have to go up and tap the wall and say, 'I hope it won't happen to mine.' " I witnessed this for myself a couple of days later. There was a news report on TV about a lost child, which distressed and agitated him. He instantly began touching his glasses (top, bottom, left, right, top, bottom, left, right), centering and recentering them in a fury. He made "whoo, whoo" noises, like an owl, and muttered sotto voce, "David, David—is *he* all right?" Then he dashed from the room to make sure. There was an intense anxiety and overconcern; an immediate alarm at the mention of any lost or hurt child; an immediate identification with himself, with his own children; an immediate, superstitious need to check up.

After tea, Bennett and I went out for a walk, past a little orchard heavy with apples and on up the hill overlooking the town, the friendly malamutes gamboling around us. As we walked, he told me something of his life. He did not know whether anyone in his family had Tourette's—he was an adopted child. His own Tourette's had started when he was about seven. "As a kid, growing up in Toronto, I wore glasses, I had bands on my teeth, *and* I twitched," he said. "That was the coup de grâce. I kept my distance. I was a loner; I'd go for long hikes by myself. I never had friends phoning all the time, like Mark—the contrast is very great." But being a loner and taking long hikes by himself toughened him as well, made him resourceful, gave him a sense of independence and self-sufficiency. He was always good with his hands and loved the structure of natural things—the way rocks formed, the way plants grew, the way animals moved, the way muscles balanced and pulled against each other, the way the body was put together. He decided very early that he wanted to be a surgeon.

Anatomy came "naturally" to him, he said, but he found medical school extremely difficult, not merely because of his tics and touchings, which became more elaborate with the years, but because of strange difficulties and obsessions that obstructed the act of reading. "I'd have to read each line many times," he said. "I'd have to line up each paragraph to get all four corners symmetrically in my visual field." Besides this lining up of each paragraph, and sometimes of each line, he was beset by the need to "balance" syllables and words, by the need to "symmetrize" the punctuation in his mind, by the need to check the frequency of a given letter, and by the need to repeat words or phrases or lines to himself.[6] All this made it impossible to read easily and fluently. Those problems are still with him and make it difficult for him to skim quickly, to get the gist, or to enjoy fine writing or narrative or poetry. But they did force him to read painstakingly and to learn his medical texts very nearly by heart.

When he got out of medical school, he indulged his interest in faraway places, particularly the North: he worked as a general practitioner in the Northwest Territories and the Yukon and worked on icebreakers circling the Arctic. He had a gift for intimacy and grew close to the Eskimos he worked with, and he became something of an expert in polar medicine. And when he married, in 1968—he was twenty-eight—he went with his bride around the world and gratified a boyhood wish to climb Kilimanjaro.

For the past seventeen years, he has practiced in small, isolated communities in western Canada—first, for twelve years, as a general practitioner in a small city. Then, five years ago, when the need to have mountains, wild country, and lakes on his doorstep grew stronger, he moved to Branford. ("And here I will stay. I never want to leave it.") Branford, he told me, has the right "feel." The people are warm but not chummy; they

---

[6]Such tendencies, common in Tourette's syndrome, are also seen in patients with postencephalitic syndromes. Thus my patient Miriam H. had compulsions to count the number of e's on every page she read; to say, or write, or spell sentences backward; to divide people's faces into juxtapositions of geometric figures; and to balance visually, to symmetrize, everything she saw.

keep a certain distance. There is a natural well-bredness and civility. The schools are of high quality, there is a community college, there are theaters and bookstores—Helen runs one of them—but there is also a strong feeling for the outdoors, for the wilds. There is much hunting and fishing, but Bennett prefers backpacking and climbing and cross-country skiing.

When Bennett first came to Branford, he was regarded, he thought, with a certain suspicion. "A surgeon who twitches! Who needs him? What next?" There were no patients at first, and he did not know if he could make it there, but gradually he won the town's affection and respect. His practice began to expand, and his colleagues, who had initially been startled and incredulous, soon came to trust and accept him, too, and to bring him fully into the medical community. "But enough said," he concluded as we returned to the house. It was almost dark now, and the lights of Branford were twinkling. "Come to the hospital tomorrow—we have a conference at seven-thirty. Then I'll do outpatients and rounds on my patients. And Friday I operate—you can scrub with me."

I slept soundly in the Bennetts' basement room that night, but in the morning I woke early, roused by a strange whirring noise in the room next to mine—the playroom. The playroom door had translucent glass panels. As I peered through them, still half-asleep, I saw what appeared to be a locomotive in motion—a large, whirring wheel going round and round and giving off puffs of smoke and occasional hoots. Bewildered, I opened the door and peeked in. Bennett, stripped to the waist, was pedaling furiously on an exercise bike while calmly smoking a large pipe. A pathology book was open before him—turned, I observed, to the chapter on neurofibromatosis. This is how he invariably begins each morning—a half hour on his bike, puffing his favorite pipe, with a pathology or surgery book open to the day's work before him. The pipe, the rhythmic exercise, calm him. There are no tics, no compulsions—at most, a little hooting. (He seems to imagine at such times that he is a prairie train.) He can read, thus calmed, without his usual obsessions and distractions.

But as soon as the rhythmic cycling stopped, a flurry of tics and compulsions took over; he kept digging at his belly, which was trim, and muttering, "Fat, fat, fat . . . fat, fat, fat . . . fat, fat, fat," and then, puzzlingly, "Fat and a quarter tit." (Sometimes the "tit" was left out.)

"What does it mean?" I asked.

"I have no idea. Nor do I know where 'Hideous' comes from—it suddenly appeared one day two years ago. It'll disappear one day, and there will be another word instead. When I'm tired, it turns into 'Gideous.' One cannot always find sense in these words; often it is just the sound that attracts me. Any odd sound, any odd name, may start repeating itself, get me going. I get hung up with a word for two or three months. Then, one morning, it's gone, and there's another one in its place." Knowing his appetite for strange words and sounds, Bennett's sons are constantly on the lookout for "odd" names—names that sound odd to an English-speaking ear, many of them foreign. They scan the papers and their books for such words, they listen to the radio and TV, and when they find a "juicy" name, they add it to a list they keep. Bennett says of this list, "It's about the most valuable thing in the house." He calls its words "candy for the mind."

This list was started six years ago, after the name Oginga Odinga, with its alliterations, got Bennett going—and now it contains more than two hundred names. Of these, twenty-two are "current"—apt to be regurgitated at any moment and chewed over, repeated, and savored internally. Of the twenty-two, the name of Slavek J. Hurka—an industrial-relations professor at the University of Saskatchewan, where Helen studied—goes the furthest back; it started to echolale itself in 1974 and has been doing so, without significant breaks, for the last seventeen years. Most words last only a few months. Some of the names (Boris Blank, Floyd Flake, Morris Gook, Lubor J. Zink) have a short, percussive quality. Others (Yelberton A. Tittle, Babaloo Mandel) are marked by euphonious polysyllabic alliterations. Echolalia freezes sounds, arrests time, preserves stimuli as "foreign bodies" or echoes in the mind, maintaining an alien existence, like implants. It is only the sound of the words, their "melody," as Bennett says, that implants them in his mind; their origins and meanings and associations are irrelevant. (There is a similarity here to his "enshrinement" of names as tics.)

"It is similar with the number compulsions," he said. "Now I have to do everything by threes or fives, but until a few months ago it was fours and sevens. Then one morning I woke up—*four* and *seven* had gone, but *three* and *five* had appeared instead. It's as if one circuit were turned on upstairs, and another turned off. It doesn't seem to have anything to do with *me*."

It is always the odd, the unusual, the salient, the caricaturable, that catch the ear and eye of the Touretter and tend to provoke elaboration and imitation.[7] This is well brought out in the personal account cited by Meige and Feindel in 1902:

> I have always been conscious of a predilection for imitation. A curious gesture or bizarre attitude affected by any one was the immediate signal for an attempt on my part at its reproduction, and is still. Similarly with words or phrases, pronunciation or intonation, I was quick to mimic any peculiarity.
>
> When I was thirteen years old I remember seeing a man

---

[7] The name of an eminent researcher on Tourette's syndrome—Dr. Abuzzahab—has an almost diagnostic power, provoking grotesque, perseverative elaborations in Touretters (Abuzzahuzzahab, etc.). The power of the unusual to excite and impress is not, of course, confined to Touretters. The anonymous author of the ancient mnemotechnic text *Ad Herennium* described it, two thousand years ago, as a natural bent of the mind and one to be exploited for fixing images more firmly in the mind:

> When we see in everyday life things that are petty, ordinary, and banal, we generally fail to remember them, because the mind is not being stirred by anything novel or marvellous. But if we see or hear something exceptionally base, dishonourable, unusual, great, unbelievable, or ridiculous, that we are likely to remember for a long time. . . . Ordinary things easily slip from the memory while the striking and the novel stay longer in the mind. . . . Let art, then, imitate nature.

with a droll grimace of eyes and mouth, and from that moment I gave myself no respite until I could imitate it accurately. . . . For several months I kept repeating the old gentleman's grimace involuntarily. I had, in short, begun to tic.

At 7:25 we drove into town. It took barely five minutes to get to the hospital, but our arrival there was more complicated than usual, because Bennett had unwittingly become notorious. He had been interviewed by a magazine a few weeks earlier, and the article had just come out. Everyone was smiling and ribbing him about it. A little embarrassed, but also enjoying it, Bennett took the joking in good part. ("I'll never live it down—I'll be a marked man now.") In the doctors' common room, Bennett was clearly very much at ease with his colleagues, and they with him. One sign of this ease, paradoxically, was that he felt free to Tourette with them—to touch or tap them gently with his fingertips or, on two occasions when he was sharing a sofa, to suddenly twist on his side and tap his colleague's shoulder with his toes—a practice I had observed in other Touretters. Bennett is somewhat cautious with his Tourettisms on first acquaintance and conceals or downplays them until he gets to know people. When he first started working at the hospital, he told me, he would skip in the corridors only after checking to be sure that no one was looking; now when he skips or hops no one gives it a second glance.

The conversations in the common room were like those in any hospitals—doctors talking among themselves about unusual cases. Bennett himself, lying half-curled on the floor, kicking and thrusting one foot in the air, described an unusual case of neurofibromatosis—a young man whom he had recently operated on. His colleagues listened attentively. The abnormality of the behavior and the complete normality of the discourse formed an extraordinary contrast. There was something bizarre about the whole scene, but it was evidently so common as to be unremarkable and no longer attracted the slightest notice. But an outsider seeing it would have been stunned.

After coffee and muffins, we repaired to the surgical-outpatients department, where half a dozen patients awaited Bennett. The first was a trail guide from Banff, very western in plaid shirt, tight jeans, and cowboy hat. His horse had fallen and rolled on top of him, and he had developed an immense pseudocyst of the pancreas. Bennett spoke with the man—who said the swelling was diminishing—and gently, smoothly palpated the fluctuant mass in his abdomen. He checked the sonograms with the radiologist—they confirmed the cyst's recession—and then came back and reassured the patient. "It's going down by itself. It's shrinking nicely—you won't be need-ing surgery after all. You can get back to riding. I'll see you in a month." And the trail guide, delighted, walked off with a jaunty step. Later, I had a word with the radiologist. "Ben-nett's not only a whiz at diagnosis," he said. "He's the most compassionate surgeon I know."

The next patient was a heavy woman with a melanoma on her buttock, which needed to be excised at some depth. Ben-nett scrubbed up, donned sterile gloves. Something about the sterile field, the prohibition, seemed to stir his Tourette's; he made sudden darting motions, or incipient motions, of his sterile, gloved right hand toward the ungloved, unwashed, "dirty" part of his left arm. The patient eyed this without ex-pression. What did she think, I wondered, of this odd darting motion, and the sudden convulsive shakings he also made with his hand? She could not have been entirely surprised, for her G.P. must have prepared her to some extent, must have said, "You need a small operation. I recommend Dr. Bennett—he's a wonderful surgeon. I have to tell you that he sometimes makes strange movements and sounds—he has a thing called Tourette's syndrome—but don't worry, it doesn't matter. It never affects his surgery."

Now, the preliminaries over, Bennett got down to the seri-ous work, swabbing the buttock with an iodine antiseptic and then injecting local anesthetic, with an absolutely steady hand. But as soon as the rhythm of action was broken for a moment—he needed more local, and the nurse held out the vial for him to refill his syringe—there was once again the darting and near-touching. The nurse did not bat an eyelid; she had seen it before and knew he would not contaminate his gloves. Now, with a firm hand, Bennett made an oval incision an inch to either side of the melanoma, and in forty seconds he had removed it, along with a Brazil-nut-shaped wodge of fat and skin. "It's out!" he said. Then, very rapidly, with great dexterity, he sewed the margins of the wound together, putting five neat knots on each nylon stitch. The patient, twisting her head, watched him as he sewed and joshed him: "Do you do all the sewing at home?"

He laughed. "Yes. All except the socks. But no one darns socks these days."

She looked again. "You're making quite a quilt."

The whole operation completed in less than three minutes, Bennett cried, "Done! Here's what we took." He held the lump of flesh before her.

"Ugh!" she exclaimed, with a shudder. "Don't show me. But thanks anyway."

All this looked highly professional from beginning to end, and, apart from the dartings and near-touchings, non-Tourettic. But I couldn't decide about Bennett's showing the excised lump to the patient. ("Here!") One may show a gallstone to a patient, but does one show a bleeding, misshapen piece of fat and flesh? Clearly, she didn't want to see it, but Bennett wanted to show it, and I wondered if this urge was part of his Tourettic scrupulosity and exactitude, his need to have everything looked at and understood. I had the same thought later in the morning, when he was seeing an old lady in whose bile duct he had inserted a T-tube. He went to great lengths to draw the tube, to explain all the anatomy, and the old lady said, "I don't want to know it. Just do it!"

Was this Bennett the Touretter being compulsive or Professor Bennett the lecturer on anatomy? (He gives weekly anatomy lectures in Calgary.) Was it simply an expression of his meticulousness and concern? An imagining, perhaps, that all patients shared his curiosity and love of detail? Some patients doubtless did, but obviously not these.

So it went on through a lengthy outpatient list. Bennett is evidently a very popular surgeon, and he saw or operated on each patient swiftly and dexterously, with an absolute and single-minded concentration, so that when they saw him they knew they had his whole attention. They forgot that they had waited, or that there were others still waiting, and felt that for him they were the only people in the world.

Very pleasant, very real, the surgeon's life, I kept thinking— direct, friendly relationships, especially clear with outpatients like this. An immediacy of relation, of work, of results, of gratification—much greater than with a physician, especially a neurologist (like me). I thought of my mother, how much she enjoyed the surgeon's life, and how I always loved sitting in at her surgical-outpatient rounds. I could not become a surgeon myself, because of an incorrigible clumsiness, but even as a child I had loved the surgeon's life, and watching surgeons at work. This love, this pleasure, half-forgotten, came back to me with great force as I observed Bennett with his patients; made me want to be more than a spectator; made me want to do something, to hold a retractor, to join in the surgery somehow.

Bennett's last patient was a young mechanic with extensive neurofibromatosis, a bizarre and sometimes cancerous disease that can produce huge brownish swellings and protruding sheets of skin, disfiguring the whole body.[8] This young man had had a huge apron of tissue hanging down from his chest, so large that he could lift it up and cover his head, and so heavy that it bowed him forward with its weight. Bennett had removed this a couple of weeks earlier—a massive procedure—with great expertise, and was now examining another huge apron descending from the shoulders, and great flaps of

[8] This was the condition, grotesquely severe, that afflicted the famous Elephant Man, John Merrick.

brownish flesh in the groins and armpits. I was relieved that he did not tic "Hideous!" as he removed the stitches from the surgery, for I feared the impact of such a word being uttered aloud, even if it was nothing but a long-standing verbal tic. But, mercifully, there was no "Hideous!"; there were no verbal tics at all, until Bennett was examining the dorsal skin flap and let fly a brief "Hid—," the end of the word omitted by a tactful apocope. This, I learned later, was not a conscious suppression—Bennett had no memory of the tic—and yet it seemed to me there must have been, if not a conscious, then a subconscious solicitude and tact at work. "Fine young man," Bennett said, as we went outside. "Not self-conscious. Nice personality, outgoing. Most people with this would lock themselves in a closet." I could not help feeling that his words could also be applied to himself. There are many people with Tourette's who become agonized and self-conscious, withdraw from the world, and lock themselves in a closet. Not so Bennett: he had struggled against this; he had come through and braved life, braved people, braved the most improbable of professions. All his patients, I think, perceive this, and it is one of the reasons they trust him so.

The man with the skin flap was the last of the outpatients, but for Bennett, immensely busy, there was only a brief break before an equally long afternoon with his inpatients on the ward. I excused myself from this to take an afternoon off and walk around the town. I wandered through Branford with the oddest sense of déjà vu and jamais vu mixed; I kept feeling that I had seen the town before, but then again that it was new to me. And then, suddenly, I had it—yes, I had seen it, I had been here before, had stopped here for a night in August 1960, when I was hitchhiking through the Rockies, to the West. It had a population then of only a few thousand and consisted of little more than a few dusty streets, motels, bars—a crossroads, little more than a truck stop in the long trek across the West. Now its population was twenty thousand, Main Street a gleaming boulevard filled with shops and cars; there was a town hall, a police station, a regional hospital, several schools—it was this that surrounded me, the overwhelming present, yet through it I saw the dusty crossroads and the bars, the Branford of thirty years before, still strangely vivid, because never updated, in my mind.

Friday is operating day for Bennett, and he was scheduled to do a mastectomy. I was eager to join him, to see him in action. Outpatients are one thing—one can always concentrate for a few minutes—but how would he conduct himself in a lengthy and difficult procedure demanding intense, unremitting concentration, not for seconds or minutes, but for hours?

Bennett preparing for the operating room was a startling sight. "You should scrub next to him," his young assistant said. "It's quite an experience." It was indeed, for what I saw in the outpatient clinic was magnified here: constant sudden dartings and reachings with the hands, almost but never quite touching his unscrubbed, unsterile shoulder, his assistant, the mirror; sudden lungings, and touchings of his colleagues with his feet; and a barrage of vocalizations—"Hooty-hooo! Hooty-hooo!"—suggestive of a huge owl.

The scrubbing over, Bennett and his assistant were gloved and gowned, and they moved to the patient, already anesthetized, on the table. They looked briefly at a mammogram on the X-ray box. Then Bennett took the knife, made a bold, clear incision—there was no hint of any ticcing or distraction—and moved straightaway into the rhythm of the operation. Twenty minutes passed, fifty, seventy, a hundred. The operation was often complex—vessels to be tied, nerves to be found—but the action was confident, smooth, moving forward at its own pace, with never the slightest hint of Tourette's. Finally, after two and a half hours of the most complex, taxing surgery, Bennett closed up, thanked everybody, yawned, and stretched. Here, then, was an entire operation without a trace of Tourette's. Not because it had been suppressed, or held in—there was never any sign of control or constraint—but because, simply, there was never any impulse to tic. "Most of the time when I'm operating, it never even crosses my mind that I have Tourette's," Bennett says. His whole identity at such times is that of a surgeon at work, and his entire psychic and neural organization becomes aligned with this, becomes active, focused, at ease, un-Tourettic. It is only if the operation is broken for a few minutes—to review a special X-ray taken during the surgery, for example—that Bennett, waiting, unoccupied, remembers that he *is* Tourettic, and in that instant he becomes so. As soon as the flow of the operation resumes, the Tourette's, the Tourettic identity, vanishes once again. Bennett's assistants, though they have known him and worked with him for years, are still astounded whenever they see this. "It's like a miracle," one of them said. "The way the Tourette's disappears." And Bennett himself was astonished, too, and quizzed me, as he peeled off his gloves, on the neurophysiology of it all.

Things were not always so easy, Bennett told me later. Occasionally, if he was bombarded by outside demands during surgery—"You have three patients waiting in the E.R.," "Mrs. X. wants to know if she can come in on the tenth," "Your wife wants you to pick up three bags of dog food"—these pressures, these distractions, would break his concentration, break the smooth and rhythmic flow. A couple of years ago, he made it a rule that he must never be disturbed while operating and must be allowed to concentrate totally on the surgery, and the O.R. has been tic-free ever since.

Bennett's operating brings up all the conundrums of Tourette's, along with deep issues such as the nature of rhythm, melody, and "flow," and the nature of acting, role, personation, and identity. A transition from uncoordinated, jerky ticciness to smoothly orchestrated, coherent movement can occur instantly in Touretters when they are exposed to, called into, rhythmic music or action. I saw this with the man I described in "Witty Ticcy Ray," who could swim the length of a pool without tics, with even, rhythmic strokes—but in the instant of turning, when the rhythm, the kinetic melody, was broken, would have a sudden flurry of tics. Many Touretters are also drawn to athletics, partly (one suspects) because of their extraordinary speed and accuracy,[9] and partly because of their bursting, inordinate motor impulse and energy, which thrust toward some motor release—but a release that, happily, instead of being explosive, can be coordinated into the flow, the rhythm, of a performance or a game.

One sees very similar situations with playing or responding to music. The convulsive or broken motor or speech patterns that may occur in Tourette's can be instantly normalized with incanting or singing (this has also long been known to occur with stutterers). It is similar with the jerky, broken movements of parkinsonism (sometimes called kinetic stutter); these too can be replaced, with music or action, by a rhythmic, melodic flow.

Such responses seem to involve chiefly the motor patterns of the individual, rather than the persona, the identity, in any higher form. *Some* of the transformation while Bennett was operating, I felt, was occurring at this elementary, "musical" level. At this level, Bennett's operating had become automatic; there

---

[9] What most of us call a startling or "abnormal" speed of movement appears perfectly normal to Touretters when they show it. This was very clear in a recent experiment of target pointing with Shane F., an artist with Tourette's. Shane showed markedly reduced reaction times, and reaching rates of almost six times normal, combined with great smoothness and accuracy of movement and aim. Such speeds were achieved quite effortlessly and naturally; normal subjects, by contrast, could achieve them, if at all, only by violent effort and with obvious compromise of accuracy and control.

On the other hand, when Shane was asked to stick to (our) normal speeds, his movements became constrained, awkward, inaccurate, and tic filled. It was clear that *his* normal and *our* normal were very different, that the Tourettic nervous system, in this sense, is more highly tuned (though, by the same token, given to precipitancy and reaction).

A similar speed and precipitancy were to be seen in many postencephalitic patients, especially when they were activated by L-DOPA. Thus, as I remarked of Hester Y., in *Awakenings*, "If Mrs. Y., before L-DOPA, was the most *impeded* person I have ever seen, she became, on L-DOPA, the most *accelerated* person I have ever seen. I have known a number of Olympic athletes, but Mrs. Y. could have beaten them all in terms of reaction time; under other circumstances she could have been the fastest gun in the West."

were, at every moment, a dozen things to attend to, but these were integrated, orchestrated, into a single seamless stream—and one that, like his driving, had become partly automated with time, so that he could chat with the nurses, make jokes, banter, think, while his hands and eyes and brain performed their skilled tasks faultlessly, almost unconsciously.

But above this level, coexisting with it, was a higher, personal one, which has to do with the identity, the role, of a surgeon. Anatomy (and then surgery) have been Bennett's constant loves, lying at the center of his being, and he is most himself, most deeply himself, when he is immersed in his work. His whole personality and demeanor—sometimes nervous and diffident—change when he puts on his surgical mantle, takes on the quiet assurance, the identity, of one who is a master at his work. It seems part of this overall change that the Tourette's vanishes, too. I have seen exactly this in Tourettic actors as well; I know one man, a character actor, who is violently Touretty offstage, but totally free from Tourettisms, totally in role, when he is acting.

Here one is seeing something at a much higher level than the merely rhythmic, quasi-automatic resonance of the motor patterns; one is seeing (however it is to be defined in psychic or neural terms) a fundamental act of incarnation or personation, whereby the skills, the feelings, the entire neural engrams of another self, are taking over in the brain, redefining the person, his whole nervous system, as long as the performance lasts.[10] Such identity transformations, reorgani-

---

[10] The matter is especially complex, for some Touretters are given to mimicry, imitation, and impersonation of a more convulsive kind. (I describe an example of this in "The Possessed.") This sort of imitation has no transformative effect; on the contrary, it thrusts the person deeper into Tourette's. The Tourettic character actor was very given to convulsive impersonations and other Tourettisms offstage, but these were quite different from the deep and healing role-playing that he was able to do onstage. The superficially imitative or impersonative impulse comes from, and stimulates, a superficial part of the person (and his neural organization)—it is only a deep, total identification, as with Bennett, that can work the transformation.

zations, occur in us all as we move, in the course of a day, from one role, one persona, to another—the parental to the professional, to the political, to the erotic, or whatever. But they are especially dramatic in those who move in and out of neurological or psychiatric syndromes, and in professional performers and actors.

These transformations, the switches between very complex neural engrams, are typically experienced in terms of "remembering" and "forgetting"—thus Bennett forgets that he is Tourettic while operating ("it never even crosses my mind"), but remembers it as soon as there is an interruption. And in the moment of remembering, he becomes so, for at this level, there is no distinction between the memory, the knowledge, the impulse, and the act—all come or go together, as one. (It is similar with other conditions: I once saw a parkinsonian man I know take a shot of apomorphine to help his rigidity and "freezing"—he suddenly unfroze a couple of minutes later, smiled, and said, "I have forgotten how to be parkinsonian.")

Friday afternoon is open. Bennett often likes to go for long hikes on Fridays, or cycle rides, or drives, with a sense of the trail, the open road, before him. There is a favorite ranch he loves to go to, with a beautiful lake and an airstrip, accessible only via a rugged dirt road. It is a wonderfully situated ranch, a narrow fertile strip perfectly placed between the lake and mountains, and we walked for miles, talking of this and that, with Bennett botanizing or geologizing as we went. Then, briefly, we went to the lake, where I took a swim; when I came out of the water I found that Bennett, rather suddenly, had curled up for a nap. He looked peaceful, tension-free, as he slept; and the suddenness and depth of his sleep made me wonder how much difficulty he encountered in the daytime, whether he might not sometimes be stressed to the limit. I wondered how much he concealed beneath his genial surface—how much, inwardly, he had to control and deal with.

Later, as we continued our ramble about the ranch, he remarked that I had seen only some of the outward expressions of his Tourette's, and these, bizarre as they occasionally seemed, were by no means the worst problems it caused him. The real problems, the inner problems, are panic and rage— feelings so violent that they threaten to overwhelm him, and so sudden that he has virtually no warning of their onset. He has only to get a parking ticket or see a police car, sometimes, for scenarios of violence to flash through his mind: mad chases, shoot-outs, flaming destructions, violent mutilation, and death scenarios that become immensely elaborated in seconds and rush through his mind with convulsive speed. One part of him, uninvolved, can watch these scenes with detachment, but another part of him is taken over and impelled to action. He can prevent himself from giving way to outbursts in public, but the strain of controlling himself is severe and exhausting. At home, in private, he can let himself go—not at others but at inanimate objects around him. There was the wall I had seen, which he had often struck in his rage, and the refrigerator, at which he had flung virtually everything in the kitchen. In his office, he had kicked a hole in the wall and had had to put a plant in front to cover it; and in his study at home the cedar walls were covered with knife marks. "It's not gentle," he said to me. "You can see it as whimsical, funny—be tempted to romanticize it—but Tourette's comes from deep down in the nervous system and the unconscious. It taps into the oldest, strongest feelings we have. Tourette's is like an epilepsy in the subcortex; when it takes over, there's just a thin line of control, a thin line of cortex, between you and it, between you and that raging storm, the blind force of the subcortex. One can see the charming things, the funny things, the creative side of Tourette's, but there's also that dark side. You have to fight it all your life."

Driving back from the ranch was a stimulating, at times terrifying, experience. Now that Bennett was getting to know me, he felt at liberty to let himself and his Tourette's go. The steering wheel was abandoned for seconds at a time—or so it seemed to me, in my alarm—while he tapped on the windshield (to a litany of "Hooty-hoo!" and "Hi, there!" and "Hideous!"), rearranged his glasses, "centered" them in a hundred different ways, and, with bent forefingers, continually smoothed and evened his mustache while gazing in the rearview mirror rather than at the road. His need to center the steering wheel in relation to his knees also grew almost frenetic: he had constantly to "balance" it, to jerk it to and fro, causing the car to zigzag erratically down the road. "Don't worry," he said when he saw my anxiety. "I know this road. I could see from way back that nothing was coming. I've never had an accident driving."[11]

The impulse to look, and to be looked at, is very striking with Bennett, and, indeed, as soon as we got back to the house he seized Mark and planted himself in front of him, smoothing his mustache furiously and saying, "Look at me! Look at me!" Mark, arrested, stayed where he was, but his eyes wandered to and fro. Now Bennett seized Mark's head, held it rigidly toward him, hissing, "Look, look at me!" And Mark became totally still, transfixed, as if hypnotized.

I found this scene disquieting. Other scenes with the family I had found rather moving: Bennett dabbing at Helen's hair, symmetrically, with outstretched fingers, going "whoo, whoo" softly. She was placid, accepting; it was a touching scene, both tender and absurd. "I love him as he is," Helen said. "I wouldn't want him any other way." Bennett feels the

---

[11] Driving cross country with another friend with Tourette's was also a memorable experience, for he would twitch the steering wheel violently from side to side, stamp on the brake or the accelerator suddenly, or pull out the ignition key at speed. But he always checked that these Tourettisms were safe, and never had an accident in ten years of driving.

same way: "Funny disease—I don't think of it as a disease but as just me. I say the word 'disease,' but it doesn't seem to be the appropriate word."

It is difficult for Bennett, and is often difficult for Touretters, to see their Tourette's as something external to themselves, because many of its tics and urges may be felt as intentional, as an integral part of the self, the personality, the will. It is quite different, by contrast, with something like parkinsonism or chorea: these have no quality of selfness or intentionality and are always felt as diseases, as outside the self. Compulsions and tics occupy an intermediate position, seeming sometimes to be an expression of one's personal will, sometimes a coercion of it by another, alien will. These ambiguities are often expressed in the terms people use. Thus the separateness of "it" and "I" is sometimes expressed by jocular personifications of the Tourette's: one Touretter I know calls his Tourette's "Toby," another "Mr. T." By contrast, a Tourettic possession of the self was vividly expressed by one young man in Utah, who wrote to me that he had a "Tourettized soul."

Though Bennett is quite prepared, even eager, to think of Tourette's in neurochemical or neurophysiological terms—he thinks in terms of chemical abnormalities, of "circuits turning on and off," and of "primitive, normally inhibited behaviors being released"—he also feels it as something that has come to be part of himself. For this reason (among others), he has found that he cannot tolerate haloperidol and similar drugs—they reduce his Tourette's, assuredly, but they reduce *him* as well, so that he no longer feels fully himself. "The side effects of haloperidol were dreadful," he said. "I was intensely restless, I couldn't stand still, my body twisted, I shuffled like a parkinsonian. It was a huge relief to get off it. On the other hand, Prozac has been a godsend for the obsessions, the rages, though it doesn't touch the tics." Prozac has indeed been a godsend for many Touretters, though some have found it to have no effect, and a few have had paradoxical effects—an intensification of their agitations, obsessions, and rages.[12]

Though Bennett has had tics since the age of seven or so, he did not identify what he had as Tourette's syndrome until he was thirty-seven. "When we were first married, he just called it a 'nervous habit,' " Helen told me. "We used to joke about it. I'd say, 'I'll quit smoking, and you quit twitching.' We thought of it as something he *could* quit if he wanted. You'd ask him, 'Why do you do it?' He'd say, 'I don't know why.' He didn't seem to be self-conscious about it. Then, in 1977, when Mark was a baby, Carl heard this program, 'Quirks and Quarks,' on the radio. He got all excited and hollered, 'Helen, come listen! This guy's talking about what I do!' He was excited to hear that other people had it. And it was a relief to me, because I had always sensed that there was something wrong. It was good to put a label on it. He never made a thing of it, he wouldn't raise the subject, but, once we knew, we'd tell people if they asked. It's only in the last few years that he's met other people with it, or gone to meetings of the Tourette Syndrome Association." (Tourette's syndrome, until very recently, was remarkably underdiagnosed and unknown, even to the medical profession, and most people diagnosed themselves, or were diagnosed by friends and family, after seeing or reading something about it in the media. Indeed, I know of another doctor, a surgeon in Louisiana, who was diagnosed by one of his own patients who had seen a Touretter on the Phil Donahue show. Even now, nine out of ten diagnoses are made, not by physicians, but by others who have learned about it from the media. Much of this media emphasis has been due to the efforts of the TSA, which had only thirty members in the early seventies but now has more than twenty thousand.)

---

[12] This was very clear with another Tourettic physician, an obstetrician, who had not only tics but panics and rages that, with a great effort, he could contain. When he was put on Prozac, this precarious control broke down, and he got into a violent fight with the police and spent a night in jail.

Saturday morning, and I have to return to New York. "I'll fly you to Calgary if the weather's fine," Bennett said suddenly last night. "Ever flown with a Touretter before?"

I had canoed with one,[13] I said, and driven across country with another, but flying with one . . .

"You'll enjoy it," Bennett said. "It'll be a novel experience. I am the world's only flying Touretter-surgeon."

When I awake, at dawn, I perceive, with mixed feelings, that the weather, though very cold, is perfect. We drive to the little airport in Branford, a veering, twitching journey that makes me nervous about the flight. "It's much easier in the air, where there's no road to keep to, and you don't have to keep your hands on the controls all the time," Bennett says. At the airport, he parks, opens a hangar, and proudly points out his airplane—a tiny red-and-white single-engine Cessna Cardinal. He pulls it out onto the tarmac and then checks it, rechecks it, and re-rechecks it before warming up the engine. It is near freezing on the airfield, and a north wind is blowing. I watch all the checks and rechecks with impatience but also with a sense of reassurance. If his Tourette's makes him check everything three or five times, so much the safer. I had a similar feeling of reassurance about his surgery—that his Tourette's, if anything, made him more meticulous, more exact, without in the least damping down his intuitiveness, his freedom.

His checking done, Bennett leaps like a trapeze artist into the plane, revs the engine while I climb in, and takes off. As we climb, the sun is rising over the Rockies to the east and floods the little cabin with a pale, golden light. We head to-

---

[13] Canoeing with Shane F. one summer on Lake Huron was a remarkable human and clinical experience, for the canoe became an extension of his body, would pitch and plunge with each of his Tourettisms, giving me an unforgettably direct sense of what it must be like to be him. We were constantly flung around, as in a storm, constantly on the point of overturning, and I longed for the canoe to founder, and sink once and for all, so that I could escape and swim back to the shore.

ward nine-thousand-foot crests, and Bennett tics, flutters, reaches, taps, touches his glasses, his mustache, the top of the cockpit. Minor tics, Little League, I think, but what if he has big tics? What if he wants to twirl the plane in midair, to hop and skip with it, to do somersaults, to loop the loop? What if he has an impulse to leap out and touch the propeller? Touretters tend to be fascinated by spinning objects; I have a vision of him lunging forward, half out the window, compulsively lunging at the propeller before us. But his tics and compulsions remain very minor, and when he takes his hands off the controls the plane continues quietly. Mercifully, there is no road to keep to. If we rise or fall or veer fifty feet, what does it matter? We have the whole sky to play with.

And Bennett, though superbly skilled, a natural aviator, *is* like a child at play. Part of Tourette's, at least, is no more than this—the release of a playful impulse normally inhibited or lost in the rest of us. The freedom, the spaciousness, obviously delight Bennett; he has a carefree, boyish look I rarely saw on the ground. Now, rising, we fly over the first peaks, the advance guard of the Rockies; yellowing larches stream beneath us. We clear the slopes by a thousand feet or more. I wonder whether Bennett, if he were by himself, might want to clear the peaks by ten feet, by inches—Touretters are sometimes addicted to close shaves. At ten thousand feet, we move in a corridor between peaks, mountains shining in the morning sun to our left, mountains silhouetted against it to our right. At eleven thousand feet, we can see the whole width of the Rockies—they are only fifty-five miles across here—and the vast golden Alberta prairie starting to the east. Every so often Bennett's right arm flashes in front of me, his hand taps lightly on the windshield. "Sedimentary rocks, look!" He gestures through the window. "Lifted up from the sea bottom at seventy to eighty degrees." He gazes at the steeply sloping rocks as at a friend; he is intensely at home with these mountains, this land. Snow lies on the sunless slopes of the mountains, none yet on their sunlit faces; and over to the northwest, toward Banff, we can see glaciers on the mountains. Bennett shifts, and shifts, and shifts again, trying to get his knees exactly symmetrical beneath the controls of the plane.

In Alberta now—we have been flying for forty minutes—the Highwood River winds beneath us. Flying due north, we start a gentle descent toward Calgary, the last, declining slopes of the Rockies all shimmering with aspen. Now, lower, to vast fields of wheat and alfalfa—farms, ranches, fertile prairie—but still, everywhere, stands of golden aspen. Beyond the checkerboard of fields, the towers of Calgary rise abruptly from the flat plain.

Suddenly, the radio crackles alive—a huge Russian air transport is coming in; the main runway, closed for maintenance, must quickly be opened up. Another massive plane, from the Zambian air force. The world's planes come to Calgary for special work and maintenance; its facilities, Bennett tells me, are some of the best in North America. In the middle of this important flurry, Bennett radios in our position and statistics (fifteen-foot-long Cardinal, with a Touretter and his neurologist) and is immediately answered, as fully and helpfully as if he were a 747. All planes, all pilots, are equal in this world. And it is a world apart, with a freemasonry of its own, its own language, codes, myths, and manners. Bennett, clearly, is part of this world and is recognized by the traffic controller and greeted cheerfully as he taxis in.

He leaps out with a startling, ticlike suddenness and celerity—I follow at a slower, "normal" pace—and starts talking with two giant young men on the tarmac, Kevin and Chuck, brothers, both fourth-generation pilots in the Rockies. They know him well. "He's just one of us," Chuck says to me. "A regular guy. Tourette's—what the hell? He's a good human being. A damn good pilot, too."

Bennett yarns with his fellow pilots and files his flight plan for the return trip to Branford. He has to return straightaway; he is due to speak at eleven to a group of nurses, and his subject, for once, is not surgery but Tourette's. His little plane is refueled and readied for the return flight. We hug and say goodbye, and as I head for my flight to New York I turn to watch him go. Bennett walks to his plane, taxis onto the main runway, and takes off, fast, with a tailwind following. I watch him for a while, and then he is gone.

To See
and Not See

Early in October of 1991, I got a phone call from a retired
minister in the Midwest, who told me about his daughter's
fiancé, a fifty-year-old man named Virgil, who had been virtually blind since early childhood. He had thick cataracts and
was also said to have retinitis pigmentosa, a hereditary condition that slowly but implacably eats away at the retinas. But
his fiancée, Amy, who required regular eye checks herself because of diabetes, had recently taken him to see her own ophthalmologist, Dr. Scott Hamlin, and he had given them new
hope. Dr. Hamlin, listening carefully to the history, was not
so sure that Virgil did have retinitis pigmentosa. It was difficult to be certain at this stage, because the retinas could no
longer be seen beneath the thick cataracts, but Virgil could
still see light and dark, the direction from which light came,
and the shadow of a hand moving in front of his eyes, so obviously there was not a total destruction of the retina. And
cataract extraction was a relatively simple procedure, done
under local anesthesia, with very little surgical risk. There
was nothing to lose—and there might be much to gain. Amy
and Virgil would be getting married soon—wouldn't it be fantastic if he could see? If, after a near-lifetime of blindness, his
first vision could be his bride, the wedding, the minister, the
church! Dr. Hamlin had agreed to operate, and the cataract on
Virgil's right eye had been removed a fortnight earlier, Amy's
father informed me. And, miraculously, the operation had

worked. Amy, who began keeping a journal the day after the operation—the day the bandages were removed—wrote in her initial entry: "Virgil can SEE! . . . Entire office in tears, first time Virgil has sight for forty years. . . . Virgil's family so excited, crying, can't believe it! . . . Miracle of sight restored incredible!" But the following day she remarked problems: "Trying to adjust to being sighted, tough to go from blindness to sighted. Has to think faster, not able to trust vision yet. . . . Like baby just learning to see, everything new, exciting, scary, unsure of what seeing means."

A neurologist's life is not systematic, like a scientist's, but it provides him with novel and unexpected situations, which can become windows, peepholes, into the intricacy of nature—an intricacy that one might not anticipate from the ordinary course of life. "Nature is nowhere accustomed more openly to display her secret mysteries," wrote William Harvey, in the seventeenth century, "than in cases where she shows traces of her workings apart from the beaten path." Certainly this phone call—about the restoration of vision in adulthood to a patient blind from early childhood—hinted of such a case. "In fact," writes the ophthalmologist Alberto Valvo, in *Sight Restoration after Long-Term Blindness*, "the number of cases of this kind over the last ten centuries known to us is not more than twenty."

What would vision be like in such a patient? Would it be "normal" from the moment vision was restored? This is what one might think at first. This is the commonsensical notion—that the eyes will be opened, the scales will fall from them, and (in the words of the New Testament) the blind man will "receive" sight.[1]

But could it be that simple? Was not *experience* necessary to see? Did one not have to learn to see? I was not well acquainted with the literature on the subject, though I had read

---

[1] There is a hint of something stranger, more complex, in Mark's description of the miracle at Bethsaida, for here, at first, the blind man saw "men as trees, walking," and only subsequently was his eyesight fully restored (Mark 8:22–26).

with fascination the great case history published in the *Quarterly Journal of Psychology* in 1963 by the psychologist Richard Gregory (with Jean G. Wallace), and I knew that such cases, hypothetical or real, had riveted the attention of philosophers and psychologists for hundreds of years. The seventeenth-century philosopher William Molyneux, whose wife was blind, posed the following question to his friend John Locke: "Suppose a man born blind, and now adult, and taught by his touch to distinguish between a cube and a sphere [be] made to see: [could he now] by his sight, before he touched them . . . distinguish and tell which was the globe and which the cube?" Locke considers this in his 1690 *Essay Concerning Human Understanding* and decides that the answer is no. In 1709, examining the problem in more detail, and the whole relation between sight and touch, in *A New Theory of Vision*, George Berkeley concluded that there was no necessary connection between a tactile world and a sight world—that a connection between them could be established only on the basis of experience.

Barely twenty years elapsed before these considerations were put to the test—when, in 1728, William Cheselden, an English surgeon, removed the cataracts from the eyes of a thirteen-year-old boy born blind. Despite his high intelligence and youth, the boy encountered profound difficulties with the simplest visual perceptions. He had no idea of distance. He had no idea of space or size. And he was bizarrely confused by drawings and paintings, by the *idea* of a two-dimensional representation of reality. As Berkeley had anticipated, he was able to make sense of what he saw only gradually and insofar as he was able to connect visual experiences with tactile ones. It had been similar with many other patients in the two hundred and fifty years since Cheselden's operation: nearly all had experienced the most profound, Lockean confusion and bewilderment.[2]

---

[2] The removal (or, as was first done, the dislocation or "couching" of the cataracted lens) leaves an eye strongly farsighted and in need of an artificial lens; and

And yet, I was informed, as soon as the bandages were removed from Virgil's eye, he saw his doctor and his fiancée, and laughed. Doubtless he saw *something*—but what did he see? What did "seeing" for this previously not-seeing man mean? What sort of world had he been launched into?

V irgil was born on a small farm in Kentucky soon after the outbreak of the Second World War. He seemed normal enough as a baby, but (his mother thought) had poor eyesight even as a toddler, sometimes bumped into things, seemed not to see them. At the age of three, he became gravely ill with a triple illness—a meningitis or meningoencephalitis (inflammation of the brain and its membranes), polio, and cat-scratch fever. During this acute illness, he had convulsions, became virtually blind, paralyzed in the legs, partly paralyzed in his breathing, and, after ten days, fell into a coma. He remained in a coma for two weeks. When he emerged from it, he seemed, according to his mother, "a different person"; he showed a curious indolence, nonchalance, passivity, seemed nothing at all like the spunky, mischievous boy he had been.

The strength in his legs came back over the next year, and his chest grew stronger, though never entirely normal. His vision also recovered significantly—but his retinas were now gravely damaged. Whether the retinal damage was caused wholly by his acute illness or perhaps partly by a congenital retinal degeneration was never clear.

In Virgil's sixth year, cataracts began to develop in both eyes, and it was evident that he was again becoming functionally blind. That same year, he was sent to a school for the blind, and there he eventually learned to read Braille and to become adept with the use of a cane. But he was not a star pu-

---

the thick lenses used in the eighteenth and nineteenth centuries, and indeed until quite recently, markedly reduced peripheral vision. Thus all patients operated upon for cataract before the present era of contact and implanted lenses had significant optical difficulties to contend with. But it was only those blind from birth or early childhood who had the special Lockean difficulty of not being able to make sense of what they saw.

pil; he was not as adventurous or aggressively independent as some blind people are. There was a striking passivity all through his time at school—as, indeed, there had been since his illness.

Yet Virgil graduated from the school and, when he was twenty, decided to leave Kentucky, to seek training, work, and a life of his own in a city in Oklahoma. He trained as a massage therapist and soon found employment at a YMCA. He was obviously good at his job, and highly esteemed, and the Y was happy to keep him on its permanent staff and to provide a small house for him across the road, where he lived with a friend, also employed at the Y. Virgil had many clients—it is fascinating to hear the tactile detail with which he can describe them—and seemed to take a real pleasure and pride in his job. Thus, in his modest way, Virgil made a life: had a steady job and an identity, was self-supporting, had friends, read Braille papers and books (though less, with the years, as Talking Books came in). He had a passion for sports, especially baseball, and loved to listen to games on the radio. He had an encyclopedic knowledge of baseball games, players, scores, statistics. On a couple of occasions, he became involved with girlfriends and would cross the city on public transport to meet them. He maintained a close tie with home, and particularly with his mother—he would get hampers of food regularly from the farm and send hampers of laundry back and forth. Life was limited, but stable in its way.

Then, in 1991, he met Amy—or, rather, they met again, for they had known each other well twenty or more years before. Amy's background was different from Virgil's: she came from a cultivated middle-class family, had gone to college in New Hampshire, and had a degree in botany. She had worked at another Y in town, as a swimming coach, and had met Virgil at a cat show in 1968. They dated a bit—she was in her early twenties, he was a few years older—but then Amy decided to go back to graduate school in Arkansas, where she met her first husband, and she and Virgil fell out of contact. She ran her own plant nursery for a while, specializing in orchids, but had to give this up when she developed severe asthma. She and her first husband divorced after a few years, and she returned to Oklahoma. In 1988, out of the blue, Virgil called her, and, after three years of long phone calls between them, they finally met again, in 1991. "All of a sudden it was like twenty years were never there," Amy said.

Meeting again, at this point in their lives, both felt a certain desire for companionship. With Amy, perhaps, this took a more active form. She saw Virgil stuck (as she perceived it) in a vegetative, dull life: going over to the Y, doing his massages; going back home, where, increasingly, he listened to ball games on the radio; going out and meeting people less and less each year. Restoring his sight, she must have felt, would, like marriage, stir him from his indolent bachelor existence and provide them both with a new life.

Virgil was passive here as in so much else. He had been sent to half a dozen specialists over the years, and they had been unanimous in declining to operate, feeling that in all probability he had no useful retinal function; and Virgil seemed to accept this with equanimity. But Amy disagreed. With Virgil being blind already, she said, there was nothing to lose, and there was a real possibility, remote but almost too exciting to contemplate, that he might actually get reasonable sight back and, after nearly forty-five years, see again. And so Amy pushed for the surgery. Virgil's mother, fearing disturbance, was strongly against it. ("He is fine as he is," she said.) Virgil himself showed no preference in the matter; he seemed happy to go along with whatever they decided.

Finally, in mid-September, the day of surgery came. Virgil's right eye had its cataract removed, and a new lens implant was inserted; then the eye was bandaged, as is customary, for twenty-four hours of recovery. The following day, the bandage was removed, and Virgil's eye was finally exposed, without cover, to the world. The moment of truth had finally come.

Or had it? The truth of the matter (as I pieced it together later), if less "miraculous" than Amy's journal suggested, was infinitely stranger. The dramatic moment stayed vacant, grew longer, sagged. No cry ("I can see!") burst from Virgil's lips. He seemed to be staring blankly, bewildered, without focusing, at the surgeon, who stood before him, still holding the bandages. Only when the surgeon spoke—saying "Well?"—did a look of recognition cross Virgil's face.

Virgil told me later that in this first moment he had no idea what he was seeing. There was light, there was movement, there was color, all mixed up, all meaningless, a blur. Then out of the blur came a voice that said, "Well?" Then, and only then, he said, did he finally realize that this chaos of light and shadow was a face—and, indeed, the face of his surgeon.

His experience was virtually identical to that of Gregory's patient S.B., who was accidentally blinded in infancy, and received a corneal transplant in his fifties:

> When the bandages were removed . . . he heard a voice coming from in front of him and to one side: he turned to the source of the sound, and saw a "blur." He realized that this must be a face. . . . He seemed to think that he would not have known that this was a face if he had not previously heard the voice and known that voices came from faces.

The rest of us, born sighted, can scarcely imagine such confusion. For we, born with a full complement of senses, and correlating these, one with the other, create a sight world from the start, a world of visual objects and concepts and meanings. When we open our eyes each morning, it is upon a world we have spent a lifetime *learning* to see. We are not given the world: we make our world through incessant experience, categorization, memory, reconnection. But when Virgil opened his eye, after being blind for forty-five years—having had little more than an infant's visual experience, and this long forgotten—there were no visual memories to support a perception; there was no world of experience and meaning awaiting him. He saw, but what he saw had no coherence. His retina and optic nerve were active, transmitting impulses, but his brain could make no sense of them; he was, as neurologists say, agnosic.

Everyone, Virgil included, expected something much simpler. A man opens his eyes, light enters and falls on the retina: he sees. It is as simple as that, we imagine. And the surgeon's own experience, like that of most ophthalmologists, had been with the removal of cataracts from patients who had almost always lost their sight late in life—and such patients do indeed, if the surgery is successful, have a virtually immediate recovery of normal vision, for they have in no sense lost their ability to see. And so, though there had been a careful surgical discussion of the operation and of possible postsurgical complications, there was little discussion or preparation for the neurological and psychological difficulties that Virgil might encounter.

With the cataract out, Virgil was able to see colors and movements, to see (but not identify) large objects and shapes, and, astonishingly, to *read* some letters on the third line of the standard Snellen eye chart—the line corresponding to a visual acuity of about 20/100 or a little better. But though his best vision was a respectable 20/80, he lacked a coherent visual field, because his central vision was poor, and it was almost impossible for the eye to fixate on targets; it kept losing them, making random searching movements, finding them, then losing them again. It was evident that the central, or macular, part of the retina, which is specialized for high acuity and fixation, was scarcely functioning, and that it was only the surrounding *para*macular area that was making possible such vision as he had. The retina itself presented a moth-eaten or piebald appearance, with areas of increased and decreased pigmentation—islets of intact or relatively intact retina alternating with areas of atrophy. The macula was degenerated and pale, and the blood vessels of the entire retina appeared narrowed.

Examination, I was told, suggested the scars or residues of

old disease but no current or active disease process; and, this being so, Virgil's vision, such as it was, could be stable for the rest of his life. It could be hoped, moreover (since the worse eye had been operated on first), that the left eye, which was to be operated upon in a few weeks' time, might have considerably more functional retina than the right.

I had not been able to go to Oklahoma straightaway—my impulse was to take the next plane after that initial phone call—but had kept myself informed of Virgil's progress over the ensuing weeks by speaking with Amy, with Virgil's mother, and, of course, with Virgil himself. I also spoke at length with Dr. Hamlin and with Richard Gregory, in England, to discuss what sort of test materials I should bring, for I myself had never seen such a case, nor did I know anyone (apart from Gregory) who had. I gathered together some materials—solid objects, pictures, cartoons, illusions, videotapes, and special perceptual tests designed by a physiologist colleague, Ralph Siegel; I phoned an ophthalmologist friend, Robert Wasserman (we had previously worked together on the case of the colorblind painter), and we started to plan a visit. It was important, we felt, not just to test Virgil but to see how he managed in real life, inside his house, outside, in natural settings and social situations; crucial, too, that we see him as a person, bringing his own life history—his particular dispositions and needs and expectations—to this critical passage; that we meet his fiancée, who had so urged the operation, and with whom his life was now so intimately mingled; that we look not merely at his eyes and perceptual powers but at the whole tenor and pattern of his life.

Virgil and Amy—now newlyweds—greeted us at the exit barrier in the airport. Virgil was of medium height, but exceedingly fat; he moved slowly and tended to cough and puff with the slightest exertion. He was not, it was evident, an entirely well man. His eyes roved to and fro, in searching movements, and when Amy introduced Bob and me he did not seem to see us straightaway—he looked toward us but not quite at us. I had the impression, momentary but strong, that he did not really look at our faces, though he smiled and laughed and listened minutely.

I was reminded of what Gregory had observed of his patient S.B.—that "he did not look at a speaker's face, and made nothing of facial expressions." Virgil's behavior was certainly not that of a sighted man, but it was not that of a blind man, either. It was, rather, the behavior of one *mentally* blind, or agnosic—able to see but not to decipher what he was seeing. He reminded me of an agnosic patient of mine, Dr. P. (the man who mistook his wife for a hat), who, instead of looking at me, taking me in, in the normal way, made sudden strange fixations—on my nose, on my right ear, down to my chin, up to my right eye—not seeing, not "getting," my face as a whole.

We walked out through the crowded airport, Amy holding Virgil's arm, guiding him, and out to the lot where they had parked their car. Virgil was fond of cars, and one of his first pleasures after surgery (as with S.B.) had been to watch them through the window of his house, to enjoy their motions, and spot their colors and shapes—their colors, especially. He was sometimes bewildered by shapes. "What cars do you see?" I asked him as we walked through the lot. He pointed to all the cars we passed. "That's a blue one, that's red—wow, that's a big one!" Some of the shapes he found very surprising. "Look at that one!" he exclaimed once. "I have to look down!" And, bending, he felt it—it was a slinky, streamlined V-12 Jaguar— and confirmed its low profile. But it was only the colors and general profiles he was getting; he would have walked past their own car had Amy not been with him. And Bob and I were struck by the fact that Virgil would look, would attend visually, only if one asked him to or pointed something out— not spontaneously. His sight might be largely restored, but using his eyes, looking, it was clear, was far from natural to him; he still had many of the habits, the behaviors, of a blind man.[3]

---

[3] One does not see, or sense, or perceive, in isolation—perception is always linked to behavior and movement, to reaching out and exploring the world. It is

To See and Not See

The drive from the airport to their house was a long one; it took us through the heart of town, and it gave us an opportunity to talk to Virgil and Amy and to observe Virgil's reactions to his new vision. He clearly enjoyed movement, watching the ever-changing spectacle through the car windows and the movement of other cars on the road. He spotted a speeder coming up very fast behind us and identified cars, buses (he especially loved the bright-yellow school buses), eighteen-wheelers, and, once, on a side road, a slow, noisy tractor. He seemed very sensitive to, and intrigued by, large neon signs and advertisements and liked picking out their letters as we passed. He had difficulty reading entire words, though he often guessed them correctly from one or two letters or from the style of the signs. Other signs he saw but could not read. He was able to see and identify the changing colors of the traffic lights as we got into town.

He and Amy told us of other things he had seen since his operation and of some of the unexpected confusions that could occur. He had seen the moon; it was larger than he expected.[4] On one occasion, he was puzzled by seeing "a fat airplane" in the sky—"stuck, not moving." It turned out to be a

---

insufficient to see; one must look as well. Though we have spoken, with Virgil, of a perceptual incapacity, or agnosia, there was, equally, a lack of capacity or impulse to *look*, to *act* seeing—a lack of visual *behavior*. Von Senden mentions the case of two children whose eyes had been bandaged from an early age, and who, when the bandages were removed at the age of five, showed no reaction to this, showed no looking, and seemed blind. One has the sense that these children, who had built up their worlds with other senses and behaviors, did not know how to *use* their eyes.

Looking—as an orientation, as a behavior—may even vanish in those who become blind late in life, despite the fact that they have been "lookers" all their lives. Many startling examples of this are given by John Hull in his autobiographical book, *Touching the Rock*. Hull had lived as a sighted man until his midforties, but within five years of becoming totally blind, he had lost the very idea of "facing" people, of "looking" at his interlocutors.

[4] Gregory's patient, too, was startled by the moon: he had expected a quarter moon would be wedge-shaped, like a piece of cake, and was astonished and amused to find it a crescent instead.

blimp. Occasionally, he had seen birds; they made him jump, sometimes, if they came too close. (Of course, they did *not* come that close, Amy explained. Virgil simply had no idea of distance.)

Much of their time recently had been spent shopping—there had been the wedding to prepare for, and Amy wanted to show Virgil off, tell his story to the clerks and shopkeepers they knew, let them see a transformed Virgil for themselves.[5] It was fun; the local television station had aired a story about Virgil's operation, and people would recognize him and come up to shake his hand. But supermarkets and other stores were also dense visual spectacles of objects of all kinds, often in bright packaging, and provided good "exercise" for Virgil's new sight. Among the first objects he had recognized, just the day after his bandages came off, were rolls of toilet paper on display. He had picked up a package and given it to Amy to prove he could see. Three days after surgery, they had gone to an IGA, and Virgil had seen shelves, fruit, cans, people, aisles, carts—so much that he got scared. "Everything ran together," he said. He needed to get out of the store and close his eyes for a bit.

He enjoyed uncluttered views, he said, of green hills and grass—especially after the overfull, overrich visual spectacles of shops—though it was difficult for him, Amy indicated, to connect the visual shapes of hills with the tangible hills he had walked up, and he had no idea of size or perspective.[6] But

---

[5] Robert Scott, a sociologist and anthropologist at the Institute for Advanced Behavioral Study at Stanford, has been especially concerned with societal reactions to the blind, and the social contempt and stigmatization so often accorded them. He has also lectured on "miracle cures," the extravagance of emotion that may attend the restoration of sight. It was Dr. Scott who, some years ago, sent me a copy of Valvo's book.

[6] Sensation itself has no "markers" for size and distance; these have to be learned on the basis of experience. Thus it has been reported that if people who have lived their entire lives in dense rain forest, with a far point no more than a few feet away, are brought into a wide, empty landscape, they may reach out and try to touch the mountaintops with their hands; they have no concept of how far the mountains are.

the first month of seeing had been predominantly positive: "Every day seems like a great adventure, seeing more for the first time each day," Amy had written, summarizing it, in her journal.

When we arrived at the house, Virgil, caneless, walked by himself up the path to the front door, pulled out his key, grasped the doorknob, unlocked the door, and opened it. This was impressive—he could never have done it at first, he said, and it was something he had been practicing since the day after surgery. It was his showpiece. But he said that in general he found walking "scary" and "confusing" without touch, without his cane, with his uncertain, unstable judgment of space and distance. Sometimes surfaces or objects would seem to loom, to be on top of him, when they were still quite a distance away; sometimes he would get confused by his own shadow (the whole concept of shadows, of objects blocking light, was puzzling to him) and would come to a stop, or trip, or try to step over it. Steps, in particular, posed a special hazard, because all he could see was a confusion, a flat surface, of parallel and crisscrossing lines; he could not see them (al-

---

Helmholtz (in *Thought in Medicine*, an autobiographical memoir) relates how, as a child of two, when walking in a park, he saw what he took to be a little tower with a rail at the top and tiny mannikins or dolls walking around behind the rail. When he asked his mother if she could reach him down one to play with, she exclaimed that the tower was a kilometer away, and two hundred meters high, and these little figures were not mannikins but *people* on the top. As soon as she said this, Helmholtz writes, he suddenly realized the scale of everything, and never again made such a perceptual mistake—though the visual perception of space as a subject never ceased to exercise him. (See Cahan, 1993.)

Poe, in "The Gold Bug," relates an opposite story: how what appeared to be a vast, many-jointed creature on a distant hill turned out to be a tiny bug on the window.

A personal experience, the first time I used marijuana, comes to mind here: gazing at my hand, seen against a blank wall. It seemed to rush away from me, while maintaining the same apparent size, until it appeared like a vast hand, a cosmic hand, across parsecs of space. Probably this illusion was made possible by, among other things, the absence of markers or context to indicate actual size and distance, and perhaps some disturbance of body image and central processing of vision.

though he knew them) as solid objects going up or coming down in three-dimensional space. Now, five weeks after surgery, he often felt more disabled than he had felt when he was blind, and he had lost the confidence, the ease of moving, that he had possessed then. But he hoped all this would sort itself out with time.

I was not so sure; every patient described in the literature had faced great difficulties after surgery in the apprehension of space and distance—for months, even years. This was the case even in Valvo's highly intelligent patient H.S., who had been normally sighted until, at fifteen, his eyes were scarred by a chemical explosion. He had become totally blind until a corneal transplant was done twenty-two years later. But following this, he encountered grave difficulties of every kind, which he recorded, minutely, on tape:

> During these first weeks [after surgery] I had no appreciation of depth or distance; street lights were luminous stains stuck to the window panes, and the corridors of the hospital were black holes. When I crossed the road the traffic terrified me, even when I was accompanied. I am very insecure while walking; indeed I am more afraid now than before the operation.

We gathered in the kitchen at the back of the house, which had a large white deal table. Bob and I laid out all our test objects—color charts, letter charts, pictures, illusions—on it and set up a video camera to record the testing. As we settled down, Virgil's cat and dog bounded in to greet and check us— and Virgil, we noted, had some difficulty telling which was which. This comic and embarrassing problem had persisted since he returned home from surgery: both animals, as it happened, were black and white, and he kept confusing them—to their annoyance—until he could touch them, too. Sometimes, Amy said, she would see him examining the cat carefully, looking at its head, its ears, its paws, its tail, and touching each part gently as he did so. I observed this myself the next day—Virgil feeling and looking at Tibbles with extraordinary intentness, correlating the cat. He would keep doing this, Amy remarked ("You'd think once was enough"), but the new ideas, the visual recognitions, kept slipping from his mind.

Cheselden described a strikingly similar scene with his young patient in the 1720s:

> One particular only, though it might appear trifling, I will relate: Having often forgot which was the cat, and which the dog, he was ashamed to ask; but catching the cat, which he knew by feeling, he was observed to look at her steadfastly, and then, setting her down, said, So, puss, I shall know you another time. . . . Upon being told what things were . . . he would carefully observe that he might know them again; and (as he said) at first learned to know, and again forgot, a thousand things in a day.

Virgil's first formal recognitions when the bandages were taken off had been of letters on the ophthalmologist's eye chart, and we decided to test him, first, on letter recognition. He could not see ordinary newsprint clearly—his acuity was still only about 20/80—but he readily perceived letters that were more than a third of an inch high. Here he did rather well, for the most part, and recognized all the commoner letters (at least, capital letters) easily—as he had been able to do from the moment the bandages were removed. How was it that he had so much difficulty recognizing faces, or the cat, and so much difficulty with shapes generally, and with size and distance, and yet so little difficulty, relatively, recognizing letters? When I asked Virgil about this, he told me that he had learned the alphabet by touch at school, where they had used letter blocks, or cutout letters, for teaching the blind. I was struck by this and reminded of Gregory's patient S.B.: "much to our surprise, he could even tell the time by means of a large clock on the wall. We were so surprised at this that we did not at first believe that he could have been in any sense blind before the operation." But in his blind days S.B. had used a large hunter watch with no glass, telling the time by touching the hands, and he had apparently made an instant "cross-modal" transfer, to use Gregory's term, from touch to vision. Virgil too, it seemed, must have been making just such a transfer.

But while Virgil could recognize individual letters easily, he could not string them together—could not read or even see words. I found this puzzling, for he said that they used not only Braille but English in raised or inscribed letters at school—and that he had learned to read fairly fluently. Indeed, he could still easily read the inscriptions on war memorials and tombstones by touch. But his eyes seemed to fix on particular letters and to be incapable of the easy movement, the scanning, that is needed to read. This was also the case with the literate H.S.:

> My first attempts at reading were painful. I could make out single letters, but it was impossible for me to make out whole words; I managed to do so only after weeks of exhausting attempts. In fact, it was impossible for me to remember all the letters together, after having read them one by one. Nor was it possible for me, during the first weeks, to count my own five fingers: I had the feeling that they were all there, but . . . it was not possible for me to pass from one to the other while counting.

Further problems became apparent as we spent the day with Virgil. He would pick up details incessantly—an angle, an edge, a color, a movement—but would not be able to synthesize them, to form a complex perception at a glance. This was one reason the cat, visually, was so puzzling: he would see a paw, the nose, the tail, an ear, but could not see all of them together, see the cat as a whole.

Amy had commented in her journal on how even the most "obvious" connections—visually and logically obvious—had to be learned. Thus, she told us, a few days after the operation "he said that trees didn't look like anything on earth," but in her entry for October 21, a month after the operation, she noted, "Virgil finally put a tree together—he now knows that the trunk and leaves go together to form a complete unit." And on another occasion: "Skyscrapers strange, cannot understand how they stay up without collapsing."

Many—or perhaps all—patients in Virgil's situation had had similar difficulties. One such patient (described by Eduard Raehlmann, in 1891), though she had had a little vision preoperatively and had frequently handled dogs, "had no idea of how the head, legs, and ears were connected to the animal." Valvo quotes his patient T.G.:

> Before the operation I had a completely different idea of space, and I knew that an object could occupy only one tactile point. I knew . . . also that if there were an obstacle or a step at the end of the porch, this obstacle occurred after a certain period of time, to which I was accustomed. After the operation, for many months, I could no longer coordinate visual sensations with my speed of walking. . . . I had to coordinate both vision and the time necessary to cover the distance. That I found very difficult. If any walking were too slow or too fast, I stumbled.

Valvo comments, "The real difficulty here is that simultaneous perception of objects is an unaccustomed way to those used to sequential perception through touch." We, with a full complement of senses, live in space and time; the blind live in a world of time alone. For the blind build their worlds from sequences of impressions (tactile, auditory, olfactory) and are not capable, as sighted people are, of a simultaneous visual perception, the making of an instantaneous visual scene. Indeed, if one can no longer see in space, then the *idea* of space becomes incomprehensible—even for highly intelligent people blinded relatively late in life (this is the central thesis of von Senden's great monograph.) And it is powerfully conveyed by John Hull in his remarkable autobiography, *Touching the Rock*, when he speaks of himself, of the blind, as "living in time" almost exclusively. With the blind, he writes,

this sense of being in a place is less pronounced. . . . Space
is reduced to one's own body, and the position of the body
is known not by what objects have been passed but by
how long it has been in motion. Position is thus measured
by time. . . . For the blind, people are not there unless
they speak. . . . People are in motion, they are temporal,
they come and they go. They come out of nothing; they
disappear.

Although Virgil could recognize letters and numbers, and
could write them, too, he mixed up some rather similar ones
("A" and "H," for example) and on occasion, wrote some
backward. (Hull describes how, after only five years of blindness in his forties, his own visual memories had become so
uncertain that he was not sure which way around a "3" went
and had to trace it in the air with his fingers. Thus the nu-
meral was retained as a tactile-motor concept, but no longer
as a visual concept.) Still, Virgil's performance was an impressive one for a man who had not seen for forty-five years. But
the world does not consist of letters and numbers. How would
he do with objects and pictures? How would he do with the
real world?
   His first impressions when the bandages were removed were
especially of color, and it seemed to be color, which has no analogue in the world of touch, that excited and delighted him—
this was very clear from the way he spoke and from Amy's
journal. (The recognition of colors and movement seems to be
innate.) It was colors to which Virgil continually alluded, the
chromatic unexpectedness of new sights. He had had Greek
salad and spaghetti the night before, he told us, and the spaghetti startled him: "White round strings, like fishing line,"
he said. "I thought it'd be brown."
   Seeing light and shape and movements, seeing colors above
all, had been completely unexpected and had had a physical
and emotional impact almost shocking, explosive. ("I felt the
violence of these sensations," wrote Valvo's patient H.S., "like
a blow on the head. The violence of the emotion . . . was akin

to the very strong emotion I felt on seeing my wife for the first time, and when out in a car, I saw the huge monuments of Rome.")

We found that Virgil easily distinguished a great array of colors and matched them without difficulty. But, confusingly, or confusedly, he sometimes gave colors the wrong names: yellow, for example, he called pink, but he knew that it was the same color as a banana. We wondered at first whether he could have a color agnosia or color anomia—defects of color association and color naming that are due to damage in specific areas of the brain. But his difficulties, it seemed to us, came simply from lack of learning (or from forgetting)—from the fact that early and long blindness had sometimes prevented his associating colors with their names or had caused him to forget some of the associations he had made. Such associations and the neural connections that underlay them, feeble in the first place, had become disestablished in his brain, not through any damage or disease, but simply from disuse.

Although Virgil believed that he had visual memories, including color memories, from the remote past—on our drive from the airport he had spoken of growing up on the farm in Kentucky ("I see the creek running down the middle," "birds on the fences," "the big old white house")—I could not decide whether these were genuine memories, visual images in his mind, or mere verbal descriptions without images (like Helen Keller's).

How was he with shapes? Here matters were more complicated, because in the weeks since his surgery Virgil had been practicing shapes, correlating their look and their feel. No such practice had been required with colors. He had at first been unable to recognize any shapes visually—even shapes as simple as a square or a circle, which he recognized instantly by touch. To him, a touch square in no sense corresponded to a sight square. This was his answer to the Molyneux question. For this reason, Amy had bought, among other things, a child's wooden formboard, with large, simple blocks—square, triangle, circle, and rectangle—to be fitted into corresponding holes, and had got Virgil to practice with it every day. Virgil found the task impossible at first, but quite easy now, after practicing for a month. He still tended to feel the holes and shapes before matching them, but when we forbade this he fitted them together quite fluently by sight alone.

Solid objects, it was evident, presented much more difficulty, because their appearance was so variable; and much of the past five weeks had been devoted to the exploration of objects, their unexpected vicissitudes of appearance as they were seen from near or far, or half-concealed, or from different places and angles.

On the day he returned home after the bandages were removed, his house and its contents were unintelligible to him, and he had to be led up the garden path, led through the house, led into each room, and introduced to each chair. Within a week, with Amy's help, he had established a canonical line—a particular line up the path, through the sitting room to the kitchen, with further lines, as necessary, to the bathroom and the bedroom. It was only from this line, at first, that he could recognize anything—though this took a great deal of interpretation and inference; thus he learned, for example, that "a whiteness to the right," to be seen as he came obliquely through the front door, was in fact the dining table in the next room, although at this point neither "table" nor "dining room" was a clear visual concept. If he deviated from the line, he would be totally disoriented. Then, carefully, with Amy's help, he started to use the line as a home base, making short sallies and excursions to either side of it, so that he could see the room, feel its walls and contents from different angles, and build up a sense of space, of solidity, of perspective.

As Virgil explored the rooms of his house, investigating, so to speak, the visual construction of the world, I was reminded of an infant moving his hand to and fro before his eyes, waggling his head, turning it this way and that, in his primal construction of the world. Most of us have no sense of the immensity of this construction, for we perform it seamlessly, unconsciously, thousands of times every day, at a glance. But this is not so for a baby, it was not so for Virgil, and it is not so for, say, an artist who wants to experience his elemental perceptions afresh and anew. Cézanne once wrote, "The same subject seen from a different angle gives a subject for study of the highest interest and so varied that I think I could be occupied for months without changing my place, simply bending more to the right or left."

We achieve perceptual constancy—the correlation of all the different appearances, the transforms of objects—very early, in the first months of life. It constitutes a huge learning task, but is achieved so smoothly, so unconsciously, that its enormous complexity is scarcely realized (though it is an achievement that even the largest supercomputers cannot begin to match). But for Virgil, with half a century of forgetting whatever visual engrams he had constructed, the learning, or relearning, of these transforms required hours of conscious and systematic exploration each day. This first month, then, saw a systematic exploration, by sight and touch, of all the smaller things in the house: fruit, vegetables, bottles, cans, cutlery, flowers, the knickknacks on the mantelpiece—turning them round and round, holding them close to him, then at arm's length, trying to synthesize their varying appearances into a sense of unitary objecthood.[7]

---

[7] There were similar problems with Gregory's subject, S.B., who never ceased to be "struck by how objects changed their shape when he walked round them. . . . He would look at a lamppost, walk round it, and stand studying it from a different aspect, and wonder why it looked different and yet the same." All newly sighted subjects, indeed, have radical difficulties with appearances, finding themselves suddenly plunged into a world that, for them, may be a chaos of continually shifting, unstable, evanescent appearances. They may find themselves completely lost, at sea, in this flux of appearances, which for them is not yet securely anchored to a world of objects, a world of space. The newly sighted, who have previously depended on senses other than vision, are baffled by the very concept of "appearance," which, being optical, has no analogue in the other senses. We who have been born into the world of appearances (and their occasional illusions, mirages, deceptions) have learned to master it, to feel secure and at home in it, but this is exceedingly difficult for the newly sighted. The philosopher F. H. Bradley wrote a famous book called *Appearance and Reality* (1893)—but for the newly sighted, at first, these have no connection.

Despite all the vexations that trying to see could entail, Virgil had stuck with this gamely, and he had learned steadily. He had little difficulty now recognizing the fruit, the bottles, the cans in the kitchen, the different flowers in the living room, and other common objects in the house.

Unfamiliar objects were much more difficult. When I took a blood-pressure cuff from my medical bag, he was completely flummoxed and could make nothing of it, but he recognized it immediately when I allowed him to touch it. Moving objects presented a special problem, for their appearance changed constantly. Even his dog, he told me, looked so different at different times that he wondered if it was the same dog.[8] He was utterly lost when it came to the rapid changes in others' physiognomies. Such difficulties are almost universal among the early blinded restored to sight. Gregory's patient S.B. could not recognize individual faces, or their expressions, a year after his eyes had been operated on, despite perfectly normal elementary vision.

What about pictures? Here I had been given conflicting reports about Virgil. He was said to love television, to follow everything on it—and, indeed, a huge new TV stood in the living room, an emblem of Virgil's new life as a seeing person. But when we tried him first on still pictures, pictures in magazines, he had no success at all. He could not see people, could not see objects—did not comprehend the idea of representation. Gregory's patient S.B. had similar problems. When shown a picture of the Cambridge Backs, showing the river and King's Bridge, Gregory tells us,

> He made nothing of this. He did not realize that the scene was of a river, and did not recognize water or bridge. . . .

---

[8] When Virgil said this I was reminded of a description in Borges's story "Funes the Memorious," where Funes's difficulty with general concepts leads him into a similar situation:

It was not only difficult for him to understand that the generic term *dog* embraced so many unlike specimens of different sizes and forms; he was disturbed by the fact that a dog at three-fourteen (seen in profile) should have the same name as the dog at three-fifteen (seen from the front).

So far as we could tell, S.B. had no idea which objects lay in front of or behind others in any of the color pictures. . . . We formed the impression that he saw little more than patches of color.

It was similar, again, with Cheselden's young patient:

We thought he soon knew what pictures represented . . . but we found afterwards we were mistaken; for about two months after he was couched, he discovered at once they represented solid bodies, when to that time he considered them only as party-coloured planes, or surfaces diversified with variety of paint; but even then he was no less surprised, expecting the pictures would feel like the things they represented, . . . and asked which was the lying sense, feeling or seeing?

Nor were things any better with moving pictures on a TV screen. Mindful of Virgil's passion for listening to baseball games, we found a channel with a game in progress. It seemed at first as if he were following it visually, because he could describe who was batting, what was going on. But as soon as we turned off the sound he was lost. It became evident that he himself perceived little beyond streaks of light and colors and motions, and that all the rest (what he *seemed* to see) was interpretation, performed swiftly, and perhaps unconsciously, in consonance with the sound. How it would be with a real game we were far from sure—it seemed possible to us that he might see and enjoy a good deal; it was in the two-dimensional representation of reality, pictorial or televisual, that he was still completely at sea.

Virgil had now had two hours of testing and was beginning to get tired—both visually and cognitively tired, as he had tended to do since the operation—and when he got tired he could see less and less, and had more and more difficulty making sense of what he could see.[9]

---

[9] Due to his exhaustion at this point, we could not test him on the visual illusions we had brought along. This was unfortunate, because "seeing" or "not

Indeed, we were getting restless ourselves and wanted to get out after a morning of testing. We asked him, as a final task before going for a drive, if he felt up to some drawing. We suggested first that he draw a hammer. (A hammer was the first object S.B. drew.) Virgil agreed and, rather shakily, began to draw. He tended to guide the pencil's movement with his free hand. ("He only does that because he's tired now," said Amy.) Then he drew a car (very high and old-fashioned); a plane (with the tail missing: it would have been hard put to fly); and a house (flat and crude, like a three-year-old's drawing).

When we finally got out, it was a brilliant October morning, and Virgil was blinded for a minute, until he put on a pair of dark-green sunglasses. Even ordinary daylight, he said, seemed far too bright for him, too glary; he felt that he saw best in quite subdued light. We asked him where he would like to go, and after thinking for a little he said, "The zoo." He had never been to a zoo, he said, and he was curious to know how the different animals looked. He had loved animals ever since his childhood days on the farm.

---

seeing" visual illusions provides an objective and replicable way of examining the visual-constructive capacities of the brain. No one has explored this approach more deeply than Gregory, and his detailed account of S.B.'s responses to visual illusions is therefore of great interest. One such illusion consists of parallel lines that, to normal eyes, seem to diverge because of the effect of diverging lines superimposed on them; no such "gestalt" effect occurred with S.B., who saw the lines as perfectly parallel—a similar lack of "influence" was seen with other illusions. Particularly interesting was S.B.'s response to reversing figures, such as cubes and staircases drawn in perspective, which are normally seen in depth and reverse their apparent configuration at intervals; the figures did not reverse for S.B. and were not seen in depth. There was, similarly, no figure-ground fluctuation with ambiguous figures. He did not, apparently, "see" distance/size changes in illusions, nor did he experience the so-called waterfall effect, the familiar aftereffect of perceived movement. In all these cases, the illusion is "seen" (even though the mind may know the perception to be illusory) by all normally sighted adults. Many of these illusory effects can also be demonstrated in young children, and some in monkeys, and even in Edelman's artificial "creature," DARWIN IV. That S.B. failed to "see" them illustrates how rudimentary his brain's powers of visual construction were, in consequence of the virtual absence of early visual experience.

Very striking, as soon as we got to the zoo, was Virgil's sensitivity to motion. He was startled, first, by an odd strutting movement; it made him smile—he had never seen anything like it. "What is it?" he asked.

"An emu."

He was not quite sure what an emu was, so we asked him to describe it to us. He had difficulty and could say only that it was about the same size as Amy—she and the emu were standing side by side at that point—but that its movements were quite different from hers. He wanted to touch it, to feel it all over. If he did that, he thought, he would then see it better. But touching, sadly, was not allowed.

His eye was caught next by a leaping motion nearby, and he immediately realized—or, rather, surmised—that it must be a kangaroo. His eye followed its motions closely, but he could not describe it, he said, unless he could feel it. We were wondering by now exactly what he could see—and what, indeed, he meant by "seeing."

In general, it seemed to us, if Virgil could identify an animal it would be either by its motion or by virtue of a single feature—thus, he might identify a kangaroo because it leapt, a giraffe by its height, or a zebra by its stripes—but he could not form any overall impression of the animal. It was also necessary that the animal be sharply defined against a background; he could not identify the elephants, despite their trunks, because they were at a considerable distance and stood against a slate-colored background.

Finally, we went to the great-ape enclosure; Virgil was curious to see the gorilla. He could not see it at all when it was half-hidden among some trees, and when it finally came into the open he thought that, though it moved differently, it looked just like a large man. Fortunately, there was a life-size bronze statue of a gorilla in the enclosure, and we told Virgil, who had been longing to touch all the animals, that he could, if nothing else, at least examine the statue. Exploring it swiftly and minutely with his hands, he had an air of assurance that he had never shown when examining anything by sight. It came to me—perhaps it came to all of us at this moment—how skillful and self-sufficient he had been as a blind man, how naturally and easily he had experienced his world with his hands, and how much we were now, so to speak, pushing him against the grain: demanding that he renounce all that came easily to him, that he sense the world in a way incredibly difficult for him, and alien.[10]

His face seemed to light up with comprehension as he felt the statue. "It's not like a man at all," he murmured. The statue examined, he opened his eyes, and turned around to the real gorilla standing before him in the enclosure. And now, in a way that would have been impossible before, he described the ape's posture, the way the knuckles touched the ground, the little bandy legs, the great canines, the huge ridge on the head, pointing to each feature as he did so. Gregory writes of a wonderful episode with his patient S.B., who had a long-standing interest in tools and machinery. Gregory took him to the Science Museum in London to see its grand collection:

> The most interesting episode was his reaction to the fine Maudeslay screw cutting lathe which is housed in a special glass case. . . . We led him to the glass case, which was closed, and asked him to tell us what was in it. He was quite unable to say anything about it, except that he thought the nearest part was a handle. . . . We then asked a museum attendant (as previously arranged) for the case to be opened, and S.B. was allowed to touch the lathe. The

---

[10] Earlier, Virgil had picked up the distant sound of lions roaring in their enclosure; he pricked up his ears and turned instantly in their direction. "Listen!" he said. "It's the lions—they're feeding the lions." The rest of us had completely missed the sound and, even when Virgil drew our attention to it, found it faint and were unsure which direction it came from. We were struck by the quality of Virgil's hearing, his auditory attention and acuteness and orientation, how extremely skilled as a listener he was. Such an acuteness and a heightening of auditory sensitivity occur in many blind people, but above all in those born blind or blinded early in life; it seems to go with the constant focusing of attention and affect and cognitive powers in these spheres, and, with this, a hyperdevelopment of auditory-cognitive systems in the brain.

result was startling.... He ran his hands eagerly over the lathe, with his eyes tight shut. Then he stood back a little and opened his eyes and said: "Now that I've felt it I can see."

So it was with Virgil and the gorilla. This spectacular example of how touching could make seeing possible explained something else that had puzzled me. Since the operation, Virgil had begun to buy toy soldiers, toy cars, toy animals, miniatures of famous buildings—an entire Lilliputian world—and to spend hours with them. It was not mere childishness or playfulness that had driven him to such pastimes. Through touching these at the same time he looked at them, he could forge a crucial correlation; he could prepare himself to see the real world by learning first to see this toy world. The disparity of scale did not matter, any more than it mattered to S.B., who was instantly able to tell the time on a large wall clock because he could correlate it with what he knew by touch from his pocket watch.

For lunch, we repaired to a local fish restaurant, and as we ate I stole glances, from time to time, at Virgil. He started eating, I observed, in the normal sighted fashion, accurately spearing segments of tomato in his salad. Then, as he continued, his aim grew worse: his fork started to miss its targets, and to hover, uncertainly, in the air. Finally, unable to "see," or make sense of, what was on his plate, he gave up the effort and started to use his hands, to eat as he used to, as a blind person eats. Amy had already told me about such relapses and described them in her journal. There had been similar reversions, for example, with his shaving, where he would start with a mirror, shaving by sight, with tense concentration. Then the strokes of the razor would become slower, and he would start to peer uncertainly at his face in the mirror, or try to confirm what he half saw by touch. Finally, he would turn away from the mirror, or close his eyes, or turn the light off, and finish the job by feel.

That Virgil should have periods of acute visual fatigue following sustained visual effort or use was scarcely surprising; all of us have them if too much is demanded of our vision. Something happens to my own visual system if, for instance, I look at EEGs nonstop for three hours: I start missing things on the traces, and seeing dazzling afterimages of the squiggles wherever I look—the walls, the ceiling, all over the visual field—and at this point I need to stop and do something else, or, even better, close my eyes for an hour. And Virgil's visual system, by comparison with the normal one, must have been at this stage labile in the extreme.

Less easy to understand, and alarming, perhaps ominous, were long periods of "blurriness"—impaired vision or gnosis—lasting hours or even days, coming on spontaneously, without obvious reason. Bob Wasserman was very much puzzled by Virgil's and Amy's descriptions of these fluctuations; he had been practicing ophthalmology for some twenty-five years and had removed many cataracts, but he had never encountered fluctuations of this sort.

After lunch, we all went to Dr. Hamlin's office. Dr. Hamlin had taken detailed photographs of the retina right after surgery, and Bob, examining the eye now (with both direct and indirect ophthalmoscopy) and comparing it with the photographs, could see no evidence of any postoperative complications. (A special test—fluorescein angiography—had shown a small degree of cystoid macular edema, but this would not have caused the rapid fluctuations that were so striking.) Because there seemed to be no adequate local or ocular cause for these fluctuations, Bob wondered whether they could be a consequence of some underlying medical condition—we had been struck by how unwell Virgil looked as soon as we met him—or whether they could represent a *neural* reaction of the brain's visual system to conditions of sensory or cognitive overload. It is no effort for the normally sighted to construct shapes, boundaries, objects, and scenes from purely visual sensations; they have been making such visual constructs, a visual world, from the moment of birth, and have developed a vast, effortless cognitive apparatus for doing so. (Normally, half of the cerebral cortex is given over to visual processing.) But in Virgil these cognitive powers, undeveloped, were rudimentary; the visual-cognitive parts of his brain might easily have been overwhelmed.

Brain systems in all animals may respond to overwhelming stimulation, or stimulation past a critical point, with a sudden shutdown.[11] Such reactions have nothing to do with the individual or his motives. They are purely local and physiological and can occur even in isolated slices of cerebral cortex: they are a biological defense against neural overload.

Still, perceptual-cognitive processes, while physiological, are also personal—it is not a world that one perceives or constructs but *one's own* world—and they lead to, are linked to, a perceptual self, with a will, an orientation, and a style of its own. This perceptual self may itself collapse with the collapse of perceptual systems, altering the orientation and the very identity of the individual. If this occurs, an individual not only becomes blind but ceases to behave as a visual being, offers no report of any change in inner state, is completely oblivious of his own visuality or lack of it. Such a condition, of total psychic blindness (known as Anton's syndrome), may occur if there is massive damage, as from a stroke, to the visual parts of the brain. But it also seemed to occur, on occasion, with Virgil. At such times, indeed, he might talk of "seeing" while in fact appearing blind and showing no visual behavior whatever. One had to wonder whether the whole basis of visual perception and identity in Virgil was as yet so feeble that under conditions of overload or exhaustion he might go in and out of not merely physical blindness but a total Anton-like psychic blindness.

A quite different sort of visual shutdown—a withdrawal—seemed to be associated with situations of great emotional

---

[11] Pavlov, speaking of such responses in dogs, called this "transmarginal inhibition consequent upon supramaximal stimulation," and regarded these shutdowns as protective in nature.

stress or conflict. And for Virgil this period was indeed as stressful a time as he had ever known: he had just had surgery, he had just been married; the even tenor of his blind, bachelor life had been shattered; he was under a tremendous pressure of expectation; and seeing itself was confusing, exhausting. These pressures had increased as his wedding day approached, especially with the convergence of his own family in town; his family had not only opposed the surgery in the first place but now insisted that he was in fact still blind. All this was documented by Amy in her journal:

October 9: Went to church to decorate for wedding. Virgil's vision quite blurry. Not able to distinguish much. It is as though sight has taken a nosedive. Virgil acting "blind" again. . . . Having me lead him around.

October 11: Virgil's family arrives today. His sight seems to have gone on vacation. . . . It is as though he has gone back to being blind! Family arrived. Couldn't believe he could see. Every time he said he could see something they would say, "Ah, you're just guessing." They treated him as though he was totally blind—leading him around, giving him anything he wanted. . . . I am very nervous, and Virgil's sight has disappeared. . . . Want to be sure we are doing the right thing.

October 12: Wedding day. Virgil very calm . . . vision little clearer, but still blurry. . . . Could see me coming down aisle, but was very blurry. . . . Wedding beautiful. Party at Mom's. Virgil surrounded by family. They still cannot accept his sight, he could not see much. Said goodbye to his family tonight. Sight began clearing up right after they left.

In these episodes Virgil was treated by his family as a blind man, his seeing identity denied or undermined, and he responded, compliantly, by acting, or even becoming, blind—a massive withdrawal or regression of part of his ego to a crushing, annihilating denial of identity. Such a regression would have to be seen as motivated, albeit unconsciously—an inhibition on a "functional" basis. Thus there seemed to be two distinct forms of "blind behavior" or "acting blind"—one a collapse of visual processing and visual identity on an organic basis (a "bottom-up" or neuropsychological disturbance, in neurological parlance), the other a collapse or inhibition of visual identity on a functional basis (a "top-down" or psychoneurotic disturbance), though no less real for him. Given the extreme organic weakness of his vision—the instability of his visual systems *and* visual identity at this point—it was very difficult, at times, to know what was going on, to distinguish between the "physiological" and "psychological." His vision was so marginal, so close to the border, that either neural overload or identity conflict might push him over it.[12]

Marius von Senden, reviewing every published case over a three-hundred-year period in his classic book *Space and Sight* (1932), concluded that every newly sighted adult sooner or later comes to a "motivation crisis"—and that not every patient gets through it. He tells of one patient who felt so threatened by sight (which would have meant his leaving the Asylum for the Blind, and his fiancée there) that he threatened to tear his eyes out; he cites case after case of patients who "behave blind" or "refuse to see" after an operation, and of others who, fearful of what sight may entail, refuse operation (one such account, entitled "L'Aveugle qui refuse de voir," was published as early as 1771). Both Gregory and Valvo dilate on the emotional dangers of forcing a new sense on a blind man—how, after an initial exhilaration, a devastating (and even lethal) depression can ensue.

Precisely such a depression descended on Gregory's patient: S.B.'s period in the hospital was full of excitement and percep-

---

[12] When a specific organic weakness exists, emotional stress can easily press toward a physical form; thus, asthmatics get asthma under stress, parkinsonians become more parkinsonian, and someone like Virgil, with borderline vision, may get pushed over the border and become (temporarily) blind. It was, therefore, exceedingly difficult at times to distinguish between what was physiological vulnerability in him, and what was "motivated behavior."

tual progress. But the promise was not fulfilled. Six months after the operation, Gregory reports,

> we formed a strong impression that his sight was to him almost entirely disappointing. It enabled him to do a little more ... but it became clear that the opportunities it afforded him were less than he had imagined. . . . He still to a great extent lived the life of a blind man, sometimes not bothering to put on the light at night. . . . He did not get on well with his neighbours [now], who regarded him as "odd," and his workmates [previously so admiring] played tricks on him and teased him for being unable to read.

His depression deepened, he became ill, and, two years after his operation, S.B. died. He had been perfectly healthy, he had once enjoyed life; he was only fifty-four.

Valvo provides us with six exemplary tales, and a profound discussion, of the feelings and behavior of early blinded people when they are confronted with the "gift" of sight and with the necessity of renouncing one world, one identity, for another.[13]

A major conflict in Virgil, as in all newly sighted people, was the uneasy relation of touch and sight—not knowing whether to feel or look. This was obvious in Virgil from the day of the operation and was very evident the day we saw

---

[13] In his ironically titled *Letter on the Blind: For the Use of Those Who Can See* (1749), the youthful Diderot maintains a position of epistemological and cultural relativism—that the blind may, in their own way, construct a complete and sufficient world, have a complete "blind identity" and no sense of disability or inadequacy, and that the "problem" of their blindness and the desire to cure this, therefore, is ours, not theirs.

He also feels that intelligence and cultivation may make a fundamental difference to what the blind may understand; may give them, at least, a formal understanding of much that they cannot directly perceive. He is especially drawn to this conclusion by pondering the case of Nicholas Saunderson, the celebrated blind mathematician and Newtonian, who died in 1740. That Saunderson, who never saw light, could conceive it so well, could be (of all things!) a lecturer in optics, could construct, in his own way, a sublime picture of the universe, excites Diderot immensely.

him, when he could hardly keep his hands off the formboard, longed to touch all the animals, and gave up spearing his food. His vocabulary, his whole sensibility, his picture of the world, were couched in tactile—or, at least, nonvisual—terms. He was, or had been until his operation, a touch person through and through.

It has been well established that in congenitally deaf people (especially if they are native signers) some of the auditory parts of the brain are reallocated for visual use. It has also been well established that in blind people who read Braille the reading finger has an exceptionally large representation in the tactile parts of the cerebral cortex. And one would suspect that the tactile (and auditory) parts of the cortex are enlarged in the blind and may even extend into what is normally the visual cortex. What remains of the visual cortex, without visual stimulation, may be largely undeveloped. It seems likely that such a differentiation of cerebral development would follow the early loss of a sense and the compensatory enhancement of other senses.

If this was the case in Virgil, what might happen if visual function was suddenly made possible, demanded? One might certainly expect *some* visual learning, some development of new pathways in the visual parts of the brain. There had never been any documentation of the kindling of activity in the visual cortex of an adult, and we hoped to take special PET scans of Virgil's visual cortex to show this as he learned to see. But what would this learning, this activation, be like? Would it be like a baby first learning to see? (This was Amy's first thought.) But the newly sighted are not on the same starting line, neurologically speaking, as babies, whose cerebral cortex is equipotential—equally ready to adapt to any form of perception. The cortex of an early blinded adult such as Virgil has already become highly adapted to organizing perceptions in time and not in space.[14]

---

[14] The Canadian psychologist Donald Hebb was deeply interested in the development of seeing and presented much experimental evidence against its being, in

An infant merely learns. This is a huge, never-ending task, but it is not one charged with irresoluble conflict. A newly sighted adult, by contrast, has to make a radical switch from a sequential to a visual-spatial mode, and such a switch flies in the face of the experience of an entire lifetime. Gregory emphasizes this, pointing out how conflict and crisis are inevitable if "the perceptual habits and strategies of a lifetime" are to be changed. Such conflicts are built into the nature of the nervous system itself, for the early blinded adult who has spent a lifetime adapting and specializing his brain must now ask his brain to reverse all this. (Moreover, the brain of an adult no longer has the plasticity of a child's brain—that is why learning new languages or new skills becomes more difficult with age. But in the case of a man previously blind, learning to see is not like learning another language; it is, as Diderot puts it, like learning language for the first time.)

In the newly sighted, learning to see demands a radical change in neurological functioning and, with it, a radical change in psychological functioning, in self, in identity. The change may be experienced in literally life-and-death terms. Valvo quotes a patient of his as saying, "One must die as a sighted person to be born again as a blind person," and the opposite is equally true: one must die as a blind person to be

---

higher animals and man, "innate," as had often been supposed. He was fascinated, understandably, by the rare "experiment" (if such a term be allowed) of restoring sight in adult life to the congenitally blind and ponders at length in *The Organization of Behaviour* on the cases collected by von Senden (Hebb himself had no personal experience of such a case). These provided rich confirmation for his thesis that seeing requires experience and learning; indeed he thought that it required, in man, fifteen years of learning to reach its full development.

But one caveat must be made (it is also made by Gregory) with regard to Hebb's comparison of the newly sighted adult to a baby. It may be that the newly sighted adult must indeed go through some of the learning and developmental stages of infancy; yet an adult, neurologically and psychologically, is nothing like a baby—an adult is already committed to a lifetime of perceptual experiences—and such cases cannot, therefore (as Hebb supposes), tell us what a baby's world is like, serve as a window into the otherwise inaccessible development of their perception.

born again as a seeing person. It is the interim, the limbo—
"between two worlds, one dead / The other powerless to be
born"—that is so terrible. Though blindness may at first be a
terrible privation and loss, it may become less so with the passage of time, for a deep adaptation, or reorientation, occurs, by
which one reconstitutes, reappropriates, the world in
nonvisual terms. It then becomes a *different* condition, a different form of being, one with its own sensibilities and coherence and feeling. John Hull calls this "deep blindness" and
sees it as "one of the orders of human being."[15]

O n October 31, the cataract in Virgil's left eye was removed, revealing a retina, an acuity, similar to the right. This
was a great disappointment, for there had been hope that it
might be a far better eye—enough to make a crucial difference
to his vision. His vision did improve slightly: he fixated better, and the searching eye movements were fewer, and he had
a larger visual field.

With both eyes working, Virgil now went back to work, but
found, increasingly, that there was another side to seeing, that
much of it was confusing, and some downright shocking. He
had worked happily at the Y for thirty years, he said, and
thought he *knew* all the bodies of his clients. Now he found
himself startled by seeing bodies, and skins, that he had previously known only by touch; he was amazed at the range of

---

[15] If blindness has a positivity of its own, is one of the orders of human being,
this is equally (or more) so for deafness, where there is not only a heightening of
visual (and, in general, spatial) abilities, but a whole community of deaf people,
with their own visuo-gestural language (Sign) and culture. Problems somewhat
similar to Virgil's may be encountered by congenitally deaf, or very early deaf-
ened, subjects given cochlear implants. Sound, for them, at first has no associations, no meaning—so they find themselves, at least initially, in a world of
auditory chaos, or agnosia. But in addition to these cognitive problems there are
identity problems, too; in a sense, they must die as deaf people to be born as hearing ones. This, potentially, is much more serious and has ramifying social and cultural implications; for deafness may be not just a personal identity, but a shared
linguistic, communal, and cultural one. These very complex issues are discussed
by Harlan Lane in *The Mask of Benevolence: Disabling the Deaf Community.*

skin colors he saw and slightly disgusted by blemishes and "stains" in skins that to his hands had seemed perfectly smooth.[16] Virgil found it a relief, when giving massages, to shut his eyes.

He continued to improve, visually, over the ensuing weeks, especially when he was free to set his own pace. He did his utmost to live the life of a sighted man, but he also became more conflicted at this time. He expressed fears, occasionally, that he would have to throw away his cane and walk outside, cross the streets, by vision alone; and, on one occasion, a fear that he might be "expected" to drive and take up an entirely new "sighted" job. This, then, was a time of great striving and real success—but success achieved, one felt, at a psychological cost, at a cost of deepening strain and splitting in himself.

There was one outing, a week before Christmas, when he and Amy went to the ballet. Virgil enjoyed *The Nutcracker*: he had always loved the music, and now, for the first time, he saw something as well. "I could see people jumping around the stage. Couldn't see what they were wearing, though," he said. He thought he would enjoy seeing a live baseball game and looked forward to the start of the season in the spring.

Christmas was a particularly festive and important time— the first Christmas after his wedding, his first Christmas as a sighted man—and he returned, with Amy, to the family farm in Kentucky. He saw his mother for the first time in more than forty years—he had scarcely been able to see her, to see anything much, at the time of the wedding—and thought she looked "real pretty." He saw again the old farmhouse, the fences, the creek in the pasture, which he had also not seen since he was a child; he had never ceased to cherish them in his mind. Some of his seeing had been a great disappointment, but seeing home and family was not—it was a pure joy.

No less important was the change in the family's attitude

---

[16] Gregory observes of S.B., "He also found some things he loved ugly (including his wife and himself!), and he was frequently upset by the blemishes and imperfections of the visible world."

toward him. "He seemed more alert," his sister said. "He would walk, move around the house, without touching the walls—he would just get up and go." She felt that there had been "a big difference" since he was first operated on, and his mother and the rest of the family felt the same.

I phoned them the day before Christmas and spoke to his mother, his sister, and others. They asked me to join them, and I wish I could have done so, for it seemed to be a joyful and affirmative time for them all. The family's initial opposition to Virgil's seeing (and perhaps to Amy, too, for having pushed it) and their disbelief that he *could* actually see had been something that he internalized, something that could literally annihilate his seeing. Now that the family was "converted," a major psychological block, one hoped, might dissolve. Christmas was the climax, but also the resolution, of an extraordinary year.

What would happen, I wondered, in the coming year? What might he hope for, at best? How much of a visual world, a visual life, might still await him? We were, frankly, quite unsure at this point. Grim and frightening though the histories of so many patients were, some, at least, overcame the worst of their difficulties and emerged into a relatively unconflicted new sight.

Valvo, normally cautious in expression, lets himself go a little in describing some of his patients' happier outcomes:

> Once our patients acquire visual patterns, and can work with them autonomously, they seem to experience great joy in visual learning . . . a renaissance of personality. . . . They start thinking about wholly new areas of experience.

"A renaissance of personality"—this was just what Amy wanted for Virgil. It was difficult for us to imagine such a renaissance in him, for he seemed so phlegmatic, so set in his ways. And yet, despite a range of problems—retinal, cortical, psychological, possibly medical—he had done remarkably well in a way, had shown a steady increase in his power to apprehend a visual world. With his predominantly positive motivation, and the obvious enjoyment and advantage he could get from seeing, there seemed no reason why he should not progress further. He could never hope to have perfect vision, but he might certainly hope for a life radically enlarged by seeing.

The catastrophe, when it came, was very sudden. On February 8, I had a phone call from Amy: Virgil had collapsed, had been taken, grey and stuporous, to the hospital. He had a lobar pneumonia, a massive consolidation of one lung, and was in the intensive-care unit, on oxygen and intravenous antibiotics.

The first antibiotics used did not work: he grew worse; he grew critical; and for some days he hovered between life and death. Then, after three weeks, the infection was finally mastered, and the lung started to reexpand. But Virgil himself remained gravely ill, for, though the pneumonia itself was clearing, it had tipped him into respiratory failure—a near-paralysis of the respiratory center in the brain, which made it unable to respond properly to levels of oxygen and carbon dioxide in the blood. The oxygen levels in his blood started to fall—fell to less than half of normal. And the level of carbon dioxide started to rise—rose to nearly three times normal. He needed oxygen constantly, but only a little could be given, lest his failing respiratory center be further depressed. With his brain deprived of oxygen and poisoned by carbon dioxide, Virgil's consciousness fluctuated and faded, and on bad days (when the oxygen in his blood was lowest and the carbon dioxide highest) he could see nothing: he was totally blind.

Much contributed to this continuing respiratory crisis: Virgil's lungs themselves were thickened and fibrotic; there was advanced bronchitis and emphysema; there was no movement of the diaphragm on one side, a consequence of his childhood polio; and, on top of all this, he was enormously obese—obese enough to cause a Pickwick syndrome (named after the somnolent fat boy, Joe, in *The Pickwick Papers*). In Pickwick syndrome, there is a grave depression of breathing, and failure to oxygenate the blood fully, associated with a depression of the respiratory center in the brain.

Virgil had probably been getting ill for some years; he had gradually been increasing in weight since 1985. But between his wedding and Christmas he had put on a further forty pounds—had shot up, in a few weeks, to two hundred and eighty pounds—partly from fluid retention caused by heart failure, and partly from nonstop eating, a habit of his under stress.

He now had to spend three weeks in the hospital, his blood oxygen still plummeting to dangerously low levels, despite his being given oxygen—and each time the level grew really low he became lethargic and totally blind. Amy would know the moment she opened his door what sort of day he was having—where the blood oxygen was—depending on whether he used his eyes, looked around, or fumbled and touched, "acted blind." (We wondered, in retrospect, whether the strange fluctuations his vision had shown from almost the day of surgery might also have been caused, at least in part, by fluctuations in his blood oxygen, with consequent retinal or cerebral anoxia. Virgil had probably had a mild Pickwick syndrome for years, and could have been close to respiratory failure and anoxia even before his acute illness.)

There was another, intermediate state, which Amy found very puzzling; at such times, he would *say* that he saw nothing whatever, but would reach for objects, avoid obstacles, and *behave* as if seeing. Amy could make nothing of this singular state, in which he manifestly responded to objects, could locate them, was seeing, and yet denied any consciousness of seeing. This condition—called implicit sight, unconscious sight, or blindsight—occurs if the visual parts of the cerebral cortex are knocked out (as they may be by a lack of oxygen, for instance), but the visual centers in the subcortex remain intact. Visual signals are perceived and are responded to appropriately, but nothing of this perception reaches consciousness at all.

At last, Virgil was able to leave the hospital and return home, but to return a respiratory cripple. He was tethered to an oxygen cylinder and could not even stir from his chair without it. It seemed unlikely at this stage that he would ever recover sufficiently to go out and work again, and the Y now felt that it had to terminate his job. A few months later, he was forced to leave the house where he had lived as an employee of the Y for more than twenty years. This was the situation that summer: Virgil had lost not only his health but his job and his house as well.

By October, however, he was feeling better and was able to go without oxygen for an hour or two at a time. It had not been wholly clear to me, from speaking to Virgil and Amy, what had finally happened to his vision after all these months. Amy said that it had "almost gone" but that now she felt it was coming back as he got better. When I phoned the visual-rehabilitation center where Virgil had been evaluated, I was given a different story. Virgil, I was told, seemed to have lost all the sight restored the previous year, with only a few bits remaining. Kathy, his therapist, thought he saw colors but little else—and sometimes colors without objects: thus he might see a haze or halo of pink around a Pepto-Bismol bottle without clearly seeing the bottle itself.[17] This color perception, she said, was the only seeing that was constant; for the rest he appeared almost blind, missed objects, groped, seemed visually lost. He was showing his old, blind random movements of the eyes. And yet sometimes, spontaneously, out of the blue, he would get sudden, startling moments of vision, in which he

---

[17] Semir Zeki has observed in some cases of cerebral anoxia that the color-constructing areas of the visual cortex may be relatively spared, so that the patient may see color *and nothing else*—no form, no boundaries, no sense of objects whatsoever.

would see objects, quite small ones. But these percepts would then vanish as suddenly as they came, and he was usually unable to retrieve them. For all practical purposes, she said, Virgil was now blind.

I was shocked and puzzled when Kathy told me this. These were phenomena radically different from anything he had shown before: What was happening now with his eyes and his brain? From a distance, I could not sort out what was happening, especially since Amy, for her part, maintained that Virgil's vision was now improving. Indeed, she got furious when she heard anyone say that Virgil was blind, and she maintained that the visual-rehab center was actually "teaching him to be blind." So in February of 1993, a year after the onset of his devastating illness, we brought Virgil and Amy to New York to see us again and to get some specialized physiological tests of retinal and brain function.

As soon as I met Virgil at the arrival gate at LaGuardia Airport, I could see for myself that everything had gone quite terribly wrong. He was now almost fifty pounds heavier than when I had met him in Oklahoma. He was carrying a cylinder of oxygen strung over one shoulder. He groped; his eyes wandered; he looked totally blind. Amy guided him, her hand under his elbow, everywhere they went. And yet sometimes as we drove over the Fifty-ninth Street Bridge into the city, he would pick up something—a light on the bridge—not guessing but seeing it quite accurately. But he could never hold it or retrieve it, and so remained visually lost.

When we came to test him in my office—first using large colored targets, then large movements and flashlights—he missed everything. He seemed totally blind—*blinder than he had been before his operations*, because then, at least, even through his cataracts he could consistently detect light, its direction, and the shadow of a hand moving before him. Now he could detect nothing whatever, no longer seemed to have any light-sensitive receptors: it was as if his retinas had gone. Yet not totally gone—that was the odd thing. For once in a while he would see something accurately: once, he saw, described, grasped, a banana; on two occasions, he was able to follow a randomly moving light bar with his hands on a computer screen; and sometimes he would reach for objects, or "guess" them correctly, even though he said he saw "nothing" at such times—the blindsight that had first been observed in the hospital.

We were dismayed at his near-uniform failure, and he was sinking into a demoralized, defeated state—it was time to stop testing and take a break for lunch. As we passed him a bowl of fruit, and he felt the fruit with swift, sensitive, skillful fingers, his face lighted up, and he regained his animation. He gave us, as he handled the fruit, remarkable tactile descriptions, speaking of the waxy, slick quality of the plum skin, the soft fuzz of peaches and smoothness of nectarines ("like a baby's cheeks"), and the rough, dimpled skin of oranges. He weighed the fruits in his hand, spoke of their weight and consistency, their pips and stones; and then, lifting them to his nose, their different smells. His tactile (and olfactory) appreciation seemed far finer than our own. We included an exceedingly clever wax pear among the real fruit; with its realistic shape and coloring, it had deceived sighted people completely. Virgil was not taken in for a moment: he burst out laughing as soon as he touched it. "It's a candle," he said immediately, somewhat puzzled. "Shaped like a bell or a pear." While he may indeed have been, in von Senden's words, "an exile from spatial reality," he was deeply at home in the world of touch, in time.

But if his sense of touch was perfectly preserved, there were, it was evident, just sparks from his retinas—rare, momentary sparks, from retinas that now seemed to be 99 percent dead. Bob Wasserman, too, who had not seen Virgil since our visit to Oklahoma, was appalled at the degradation of vision and wanted to reexamine the retinas. When he did so, they looked exactly as before—piebald, with areas of increased and decreased pigmentation. There was no evidence of any new disease. Yet the functioning of even the preserved areas of retina had fallen to almost zero. Electroretinograms, designed to record the retina's electrical activity when stimulated by light, were completely flat, and visual evoked potentials, designed to show activity in the visual parts of the brain, were absent, too—there was no longer anything, electrically, going on in either the retinas or the brain that could be recorded. (There may have been rare, momentary sparks of activity, but if so, we failed to catch these in our recordings.) This inactivity could not be attributed to the original disease, retinitis, which had long been inactive. Something else had emerged in the past year and had, in effect, extinguished his remaining retinal function.

We remembered how Virgil had constantly complained of glare, even on relatively dull, overcast days—how glare seemed to blind him sometimes, so that he needed the darkest glasses. Was it possible (as my friend Kevin Halligan suggested) that with the removal of his cataracts—cataracts that had perhaps shielded his fragile retinas for decades—the ordinary light of day had proved lethal, burnt out his retinas? It is said that patients with other retinal problems, like macular degeneration, may be exceedingly intolerant of light—not merely ultraviolet but light of all wavelengths—and that light may hasten the degeneration of their retinas. Was this what had happened with Virgil? It was one possibility. Should we have foreseen it and rationed Virgil's sight, or the ambient light, in some way?

Another possibility—a likelier one—related to Virgil's continuing hypoxia, the fact that he had not had properly oxygenated blood for a year. We had clear accounts of his vision waxing and waning in the hospital as his blood gases went up and down. Could the repeated, or continuing, oxygen-starving of his retinas (and perhaps also of the visual areas of his cortex) have been the factor that did them in? It was wondered, at this point, whether raising blood oxygenation to 100 percent (which would have required sustained artificial respiration with pure oxygen) might restore some retinal or cerebral function. But it was decided that this procedure would be too risky, since it might cause long-term or permanent depression of the brain's respiratory center.

This, then, is Virgil's story, the story of a "miraculous" restoration of sight to a blind man, a story basically similar to that of Cheselden's young patient in 1728, and of a handful of others over the past three centuries—but with a bizarre and ironic twist at the end. Gregory's patient, so well adapted to blindness before his operation, was first delighted with seeing, but soon encountered intolerable stresses and difficulties, found the "gift" transformed to a curse, became deeply depressed, and soon after died. Almost all the earlier patients, indeed, after their initial euphoria, were overwhelmed by the enormous difficulties of adapting to a new sense, though a very few, as Valvo stresses, have adapted and done well. Could Virgil have surmounted these difficulties and adapted to seeing where so many others had foundered on the way?

We shall never know, for the business of adaptation—and, indeed, of life as he knew it—was suddenly cut across by a gratuitous blow of fate: an illness that, at a single stroke, deprived him of job, house, health, and independence, leaving him a gravely sick man, unable to fend for himself. For Amy, who incited the surgery in the first place, and who was so passionately invested in Virgil's seeing, it was a miracle that misfired, a calamity. Virgil, for his part, maintains philosophically, "These things happen." But he has been shattered by this blow, has given vent to outbursts of rage: rage at his helplessness and sickness; rage at the smashing of a promise and a dream; and beneath this, most fundamental of all, a rage that had been smoldering in him almost from the beginning—rage at being thrust into a battle he could neither renounce nor win. At the beginning, there was certainly amazement, wonder, and sometimes joy. There was also, of course, great courage. It was an adventure, an excursion into a new world, the like of which is given to few. But then came the problems, the conflicts, of seeing but not seeing, not being able to make a visual world, and at the same time being forced to give up his own. He found himself between two worlds, at home in neither—a torment from which no escape seemed possible. But then, paradoxically, a release was given, in the form of a second and now final blindness—a blindness he received as a gift. Now, at last, Virgil is allowed to not see, allowed to escape from the glaring, confusing world of sight and space, and to return to his own true being, the intimate, concentrated world of the other senses that had been his home for almost fifty years.

The Landscape of
His Dreams

I first met Franco Magnani in the summer of 1988, when the Exploratorium in San Francisco held a symposium and an exhibit on memory. The exhibit included fifty paintings and drawings by him—all of Pontito, the little Tuscan hill town where he was born but had not seen for more than thirty years. Next to them, in astounding apposition, were photographs of Pontito taken by the Exploratorium's photographer, Susan Schwartzenberg, from exactly the same viewpoints as Magnani's, wherever possible. (This was not always possible, because Magnani sometimes visualized and painted Pontito from an imaginary aerial viewpoint fifty or five hundred feet above the ground; Schwartzenberg sometimes had to hoist her camera aloft on a pole and at one point thought of hiring a helicopter or a balloon.) Magnani was billed as "A Memory Artist," and one had only to glance at the exhibit to see that he indeed possessed a prodigious memory—a memory that could seemingly reproduce with almost photographic accuracy every building, every street, every stone of Pontito, far away, close up, from any possible angle. It was as if Magnani held in his head an infinitely detailed three-dimensional model of his village, which he could turn around and examine, or explore mentally, and then reproduce on canvas with total fidelity.

My first thought when I saw the resemblance between the paintings and the photographs was that here was that rare phenomenon, an eidetic artist: an artist able to hold in memory, for hours or days (perhaps for years), an entire scene that has been glimpsed in a flash; the commander (or slave) of a prodigious native power of imagery and memory. But an eidetic artist would scarcely confine himself to a single theme or subject; on the contrary, he would exploit his memory, or display it, in a huge range of subjects, to show that nothing lay beyond its grasp—whereas Magnani seemingly wanted to concentrate it exclusively upon Pontito. This, then, was an exhibit not of "pure" memory but of memory harnessed to a single, overwhelming motive: the recollection of his childhood village. And, I now realized, it was not just an exercise in memory; it was, equally, an exercise in nostalgia—and not just an exercise but a compulsion, and an art.

A few days later, I spoke to Franco and arranged to meet him at his house. He lives in a small community a few miles outside San Francisco. Once I had found his street, I did not need to look for his house number, because his house stood out immediately from its neighbors. In the small front yard was a low stone wall, resembling those in his paintings of Pontito; his car, an aging sedan with vanity plates ("Pontito"), was parked in the street; the garage had been converted into a studio, and its door was wide open, revealing the artist himself, intently at work.

Franco was tall and slim, with enormous horn-rimmed glasses that magnified his eyes. He had thick brown hair, carefully parted on one side; a springy stride; and an air of great exuberance and vitality—he was fifty-four but seemed much younger. He invited me in and showed me around his home. Every room had paintings on every wall, and every drawer and closet seemed stacked full of paintings—it was less like a house than a museum or archive, totally devoted to the recollection, the reproduction, of Pontito.

As we walked through the house, each painting arrested his attention, aroused a flood of reminiscence: what happened here, what there, and how so-and-so stood there once. "Look at this wall here—that's where the priest, he caught me climbing into the garden behind the church. He chase me all the way down the street. Oh, he always chase all the kids away from there." Each reminiscence triggered others, and these still others, so that within minutes we were engulfed in a flood, without any clear direction or center, but all relating to his early life—to Pontito as he had experienced it as a child. He leapt from one story to another, without any connection that I could discern. This sort of rambling—single-minded and intense but incoherent and unfocused—seemed characteristic of Franco: it showed the quality of his obsession, the fact that he thought of Pontito day and night, to the exclusion of all else.

As Franco talked, I had the impression that his reminiscences were taking him over, that these upsurging memories drove him, dominated him, exerted a huge, irresistible force. He would gesture; he would mime; he would breathe heavily; he would glare—he seemed to be completely transported. Then, with a start, he would come back, smile a little embarrassedly, and say, "That's how it was."

This nonstop verbosity, this reminiscence of concrete episodes, seemed to be in a quite different mode from his painting. When he was alone, he said, the yammer and clatter of memories would die down, and he would get a calm impression of Pontito: a Pontito without people, without incidents, without temporality; a Pontito at peace, suspended in a timeless "once," the "once" of allegory, fantasy, myth, and fairy tale.

By midmorning, I had been enthralled again by Franco's paintings but had had enough of his reminiscences. He had one subject only—could talk of nothing else. What could be more sterile, more boring? Yet out of this obsession he could create a lovely, real, and tranquil art. What was it that served to transform his memories—to remove them from the sphere of the personal, the trivial, the temporal, and bring them into the realm of the universal, the sacred? One encounters boring talkers, reminiscers, by the score, and not one of them will be a true artist, like Franco. Thus it was not just his vast memory or his obsession that was crucial in making him an artist but, rather, something much deeper.

Franco was born in Pontito in 1934. A village of some five hundred souls, it was nestled in the hills of Castelvecchio, in the province of Pistoia, about forty miles west of Florence. Like all Tuscan hill villages, it had an ancient lineage and still had an abundance of Etruscan tombs, as well as traditional patterns of farming, terracing, and olive and vine growing, going back more than two thousand years. Its stone buildings, its steep, winding streets, traversable only by trim mountain donkeys or human feet, had not changed in centuries, nor had the simple, orderly life of its residents. The village was dominated, at its highest point, by the spire of its ancient church, and Franco's house was next to the church—indeed, as a child, he could nearly touch its roof if he leaned far enough out of his bedroom window. Somewhat isolated and inbred, the villagers formed almost a single large family: the Magnanis, the Papis, the Vanuccis, the Tamburis, the Sarpis, were all related. The village's greatest eminence was Lazzaro Papi, an eighteenth-century commentator on the French Revolution; a monument to him still stands in the central square.

Isolated, unchanging, traditional, Pontito was a citadel against the flux of change and time. The earth was fertile, the inhabitants industrious; their farms and orchards sustained them without luxury or want. Life was good, and secure, for Franco, for all the villagers, until the outbreak of the war.

But then came horrors and troubles of every kind. Franco's father died in an accident in 1942, and the following year saw the entry of the Nazis, who took over the village and evicted the townspeople. When the villagers came back, many of their houses had been defaced. Life was never the same after this. The town had been despoiled, the fields and the orchards had been ruined, and, perhaps most important, the old patterns and mores disturbed. Pontito gathered itself together and tried gallantly to recoup after the war, but it failed to recover fully.

It has been in a slow decline ever since. Its orchards and fields, its agrarian economy, were never fully restored; it ceased to be self-sufficient economically, and its young men and women had to leave and go elsewhere. The once-thriving village, which had five hundred people before the war, has only seventy people now, all elderly or retired. There are no longer any children, and there are few working adults. The once-vital town is depopulated and dying.

All of Franco's paintings represent Pontito, and his life there, prior to 1943; they are all recollections of his childhood, of the place where he lived and played and grew up, before his father was killed, before the Germans came, before the occupation of the village and the ruination of its fields.

Franco lived in Pontito until he was twelve, in 1946, when he went to school in Lucca. In 1949, he went on to Montepulciano, as an apprentice furniture maker. He was remarkable for his "photographic" memory even before this (as were his mother and one of his sisters, to a lesser degree): he could remember a page after a single reading or the lesson in church after a single hearing; he could remember all the inscriptions on the gravestones in the cemetery; he could remember (and add up) long lists of figures at a glance. But it was only in Lucca, away from home for the first time, and markedly homesick, that he started to experience another sort of memory: images that darted suddenly into his mind—images of great personal resonance and intensity, sharp with pleasure or pain. These images were quite different from the "rote" memory that had distinguished him thus far; they were involuntary and sudden, flashlike and imperative—hallucinatory, almost, in their sound, their texture, their smell, and their feel. This new kind of memory was, above all, experiential or autobiographical, for every image came with its proper personal context and affect. Each image was a scene, a flashback, from his life. "He painfully missed Pontito," his sister told me. "It was the church, the street, the fields, that he would 'see'—but as yet he had no impulse to draw."

Franco returned to Pontito in 1953, after his four years of

apprenticeship, but found that the village, already declining, could not support a woodworker. Unable to make a living in Pontito, or follow his trade, he went to Rapallo, where he worked as a cook—though he remained dissatisfied and dreamed of a different life and faraway places. At the start of 1960—he was now twenty-five—he decided, half impulsively, half deliberately, to quit his job, to see the world, to work as a cook on a cruise ship. And as he was preparing to do this (knowing, perhaps, that he would never return) he composed an autobiography—but he flung it into the water as he was boarding the ship. The need to recollect, to make a picture of his childhood, was clearly very strong in him at this point; but he had not yet found his medium. So he set sail. He plied to and fro between the Caribbean and Europe and got to know Haiti, the Antilles, and the Bahamas well—indeed, in 1963 and '64 he spent fourteen months in Nassau. During this time, he says he "forgot" Pontito—thoughts of it almost never came to his mind.

In 1965, when he was thirty-one, he made a momentous decision: he would not go back to Italy, not go back to Pontito; he would settle in America, in San Francisco. The decision was a difficult and troubling one. It threatened a separation, perhaps irrevocable, from all that he held most valuable and dear: his country, his language, his village, his family, the habits and traditions that had bound his people together for hundreds of years. But it promised, or seemed to promise, freedom and perhaps wealth, a new life in a new country, a freedom to be himself, to be independent, such as he had tasted on board ship. (His father, as a young man, had also gone to America and was in business for a few years, but then languished, and returned to Pontito.)

But with the troubling decision a strange illness occurred, which finally brought him to a sanatorium. It is far from clear what the illness was. There was a crisis of decision, and hope and fear, but there was also a high fever, weight loss, delirium, perhaps seizures; there was a suggestion of tuberculosis, of a psychosis, or of some neurological condition. It was never really resolved what was going on, and the nature of the illness remains a mystery. What is clear is that at the height of the illness, his brain perhaps stimulated by excitement and fever, Franco started to have, nightly and all night, overwhelmingly vivid dreams. Every night, he dreamed of Pontito, not of his family, not of activities or events, but of the streets, the houses, the masonry, the stones—dreams with the most microscopic, veridical detail, a detail beyond anything he could consciously remember. An intense, strange excitement possessed him in these dreams: a sense that something had just happened, or was about to; a sense of immense, portentous, yet enigmatic significance, accompanied by an insatiable, yearning, bittersweet nostalgia. And when he awoke it seemed to him he was not fully awake, for the dreams were still present, still before his inward eye, painting themselves on the bedclothes and the ceiling and the walls all around him, or standing on the floor, like models, solid with exquisite detail.

In the hospital, with these dreamlike images forcing themselves upon his consciousness and his will, a new feeling took hold of him—a sense that he was now being "called." Though his powers of imagery had always been great, he had never seen images of such intensity before—images that suspended themselves like apparitions in the air and promised him a "repossession" of Pontito. Now they seemed to say to him, "Paint us. Make us real."

What happened, one wonders (and Franco has never ceased to ask himself), in those days and nights in the hospital, that time of crisis, delirium, fever, seizures? Did he crack under the stress of his decision, undergo a "Freudian" splitting of the ego, and become from then on a sort of hypermnesic hysteric? ("Hysterics suffer mainly from reminiscences," Freud wrote.) Did a split-off part of him seek to provide in memory or fantasy what he had cut himself off from, could no longer return to in reality? Were these dreams, these memory images, called up by him in response to a deep emotional need? Or were they forced on him by some strange, physiological bombardment of the brain, a process that he (as a person) had nothing to do with, but could not help reacting to? Franco considered, but rejected, these "medical" possibilities (and never allowed them to be properly explored) and moved instead to a more spiritual one.[1] A gift, a destiny, had been vouchsafed to him, he felt, and it was his task to obey, not to question. It was in this religious spirit, then, that Franco, after a brief struggle, accepted his visions and now dedicated himself to making them a palpable reality.

Though he had scarcely painted or drawn before, he felt he could take a pen or brush and trace the outlines that hovered so clearly in the air before him or projected themselves, as through a camera lucida, on the white walls of his room. Above all, in those first nights of crisis, there came to him images of the house where he was born, images impossibly beautiful but with a menacing aspect, too.

Franco's first Pontito painting, indeed, was of his house, a painting that, despite his lack of training, had a striking confidence and clarity of outline, and a strange, dark emotional force. Franco himself was amazed by this painting, by the fact that he could paint, could express himself in this wonderful new way. Even now, a quarter of a century later, he remains amazed. "Fantastic," he says. "Fantastic. How could I do it? And how could I have had the gift and not known it before?" He had occasionally, as a child, imagined himself as an artist, but that was a mere fancy, and he had never done more than play with a pen or a brush—sketch a ship on a postcard, perhaps, or a Caribbean scene. He was also frightened by the power he now felt—a power that had seized him and taken him over but that he could perhaps control and give voice to.

---

[1] Giorgio De Chirico, the painter, was subject to classical migraines and migraine auras of great severity—he gave vivid circumstantial accounts of these—and sometimes incorporated their geometric patterns, their zigzags, their blinding lights and darkness, into his pictures (this has been described in detail by G. N. Fuller and M. V. Gale in the *British Medical Journal*). But De Chirico was also reluctant to acknowledge a purely medical or physical cause for his visions, since, he felt, their spiritual quality was so strong. His final term for them was a compromise—"spiritual fevers."

And the voice of his paintings, his style, was there from the start, even—or especially—in the first paintings he did. "The first two paintings are quite different from the later ones," his friend Bob Miller said to me. "There's something ominous in them—you can see something deep and significant happening."

That Franco did not start thinking obsessively of Pontito— did not dream day and night of Pontito—until this time is corroborated by his brother-in-law, who did not see him between 1961 and 1987. "Back in '61, Franco would talk about anything," he told me. "He wasn't obsessed—he was normal. But when I saw him in '87 he seemed possessed. He constantly had visions of Pontito, and he wouldn't talk about anything else."

Miller says, "His paintings started in this crisis period. He was in the hospital, pretty near a mental breakdown, and the paintings seemed to be a sort of solution, or cure. Sometimes he says, 'I have these memories, I have these dreams, I can't function,' but he seems to function pretty well. It's hard to have a normal conversation with him, though—it's 'Pontito, Pontito, Pontito,' all the time. It's as if he had this 3-D construct, this model of Pontito, he can erect—he moves his head, turns around, to 'see' different aspects. He seemed to think this sort of 'seeing' was normal, and it was only in the late seventies, when Gigi, a friend, came back with photos of Pontito, that he realized for the first time how extraordinary it was. . . . Everything is fresh, excited, as if just recalled. It is not a fixed thing, a repertorial thing, at all. He remembers scenes. He acts them out, relives the whole thing. So it is a very concrete, particular memory, which organizes itself in stories and scenes—a memory of who said what when." One sometimes feels that there is something theatrical in the paintings, and, to some extent, Franco himself sees them that way.

The mood that had announced itself in dreams at night now deepened and intensified in Franco's mind. He started to get "visions" of Pontito by day—visions emotionally overwhelming but with a minute and three-dimensional quality that he compares to holography. These visions may come at any time—when he is eating or drinking, taking a stroll, showering. There is no doubt of their reality for him. He is, perhaps, talking with you quietly, and suddenly he leans forward, his eyes fixed and staring, in a rapture: an apparition of Pontito is rising before him. "Many of Franco's paintings," writes Michael Pearce (in a fascinating analysis that appeared in the *Exploratorium Quarterly* to coincide with the exhibition), "begin with what he describes as a kind of memory flash, where a particular scene will suddenly come into his head. He often feels a great urgency to get the scene down on paper immediately, and has been known to leave a bar in mid-drink in order to begin a sketch.... Apparently the 'flash' Franco gets of a scene is not a static, photographic view.... He can scan the area and 'see' in several directions. To do this, he must physically reorient his body, turning to the right to envision what would be to the right in the Pontitan scene, to the left to 'see' to that side ... his eyes looking into the distance as though he can see the stone buildings and archways and streets."

Such apparitions are not only visual. Franco can hear the church bells ("like I was there"); he can feel the churchyard wall; and, above all, he can smell what he sees—the ivy on the church wall, the mingled smells of incense, must, and damp, and, admixed with these, the faint smell of the nut and olive groves that grew around the Pontito of his youth. Sight, sound, touch, smell, at such times are almost inseparable for Franco, and what comes to him is like the complex and coenesthetic experiences of early childhood—"the instantaneous records of total situations," the psychiatrist Harry Stack Sullivan once called them.

It seems likely that there is some sudden and profound change in Franco's brain whenever he is "inspired" or "possessed." Certainly when I first saw Franco seized by a vision, and noted his staring eyes, his dilated pupils, his raptus of attention, I could not help wondering whether he was having a sort of psychic seizure. Such psychic seizures were first recognized a century ago by the great neurologist John Hughlings Jackson, who stressed the commanding hallucinations, the flow of involuntary "reminiscence," the sense of revelation, and the strange, half-mystical "dreamy state" that could be characteristic of these. Such seizures are associated with epileptic activity in the temporal lobes of the brain.

In the last century, Hughlings Jackson, among others, suspected that some patients with frequent psychical seizures might show strange alterations in thinking and personality with the onset of their disorder. But it was not until the 1950s and 1960s that such an "interictal personality syndrome," as it came to be called, received closer attention. In 1956 the French neurologist Henri Gastaut wrote an important memoir on van Gogh, in which he presented the case for van Gogh having not only temporal lobe seizures but a characteristic personality change with the onset of these, gradually intensifying for the rest of his life. In 1961 one of the most gifted of American neurologists, Norman Geschwind, spoke about the possible role of temporal lobe epilepsy in Dostoevsky's life and writings, and by the early seventies had become convinced that a number of patients with TLE showed a peculiar intensification (but also narrowing) of emotional life, "an increased concern with philosophical, religious and cosmic matters." Remarkable productiveness was seen in many patients: the writing of autobiographies, the filling of endless diaries, obsessive drawing (in those graphically inclined)—and a general sense of illumination, "mission," and "fate," this even in poorly educated, "unintellectual" people who had shown no dispositions in these directions before.

Geschwind's first publications regarding the incidence and nature of the syndrome were published in 1974 and 1975, with his colleague Stephen Waxman, and galvanized the neurological world. Here, for the first time, a whole constellation of symptoms and behaviors traditionally suggestive of either mental illness or inspiration were attributed to a specific neurological cause, in particular (as David Bear, another colleague, was to stress) a "hyperconnectivity" between the sensory and emotional parts of the brain, resulting in greatly heightened and emotionally charged perceptions, memories, and images. "Personality change in temporal lobe epilepsy," Geschwind observed, "may well be the most important single set of clues we possess to deciphering the neurological systems that underlie the emotional forces that guide behavior."

Such changes, Geschwind emphasized, could not be considered either negative or positive as such; what mattered was the role they came to play in a person's life, and this could be creative or destructive, adaptive or maladaptive. He was, however, especially interested in the (relatively uncommon) situation of a highly creative use of the syndrome. "When this tragic disease is visited upon a man of genius," he wrote of Dostoevsky, "he is able to extract from it a depth of understanding . . . a deepening of emotional response."[2] It was the conjunction of disease, or biological disposition, with individual creativity that excited Geschwind above all.[3]

---

[2] This, too, was Dostoevsky's attitude. "What if it is disease," he asks, through Prince Mishkin. "What does it matter that it is an abnormal intensity, if the result, if the minute of sensation, remembered and analysed afterwards in health, turns out to be the acme of harmony and beauty . . . of completeness, of proportion?"

[3] Although the interpretation of the lives and works and personalities of eminent figures in terms of their supposed neurological or psychiatric dispositions is not new, it has become an obsession, almost an industry, at the present time. Eve LaPlante, in her book *Seized*, speaks of the characteristic "marks" of TLE and Geschwind syndrome not only in van Gogh and Dostoevsky, but in figures as various as Poe, Tennyson, Flaubert, Maupassant, Kierkegaard, and Lewis Carroll (to say nothing of such contemporaries as Walker Percy, Philip Dick, and Arthur Inman of the 155-volume diary). William Gordon Lennox (author of a massive 2-volume standard work on epilepsy) adds scores of others to this list, from Socrates, Paul, and the Buddha to Newton, Strindberg, Rasputin, Paganini, and Proust. The famous, sudden returns of memory in *The Remembrance of Things Past* are all seen by Lennox as hypermnesic or experiential seizures, brought on by particular stimuli evocative of the past.

Other books and articles attribute Tourette's syndrome to Samuel Johnson and Mozart, autism to Bartók and Einstein, and manic-depressive illness to virtually

The rather dry term "interictal personality syndrome" was to become "Waxman-Geschwind syndrome," or sometimes simply "Dostoevsky syndrome." I had to wonder whether the illness that Franco had in 1965, with its intensely vivid dreams, its seizurelike hallucinations, its mystical illuminations and transports, was not indeed the inauguration of such a Dostoevsky syndrome.

Hughlings Jackson speaks of the "doubling of consciousness" that tends to occur in such seizures. And this is how it is with Franco: when he is seized by a vision, a waking dream, a reminiscence of Pontito, he is transported—he is, in a sense, there. His reminiscences come suddenly, unannounced, with the force of revelation. Though he has learned over the years to control them to some extent, to invoke them or conjure them up—as indeed all artists learn to do—they remain essentially involuntary. It is precisely this characteristic that Proust holds to be the most valuable: to his mind, voluntary recall is conceptual, conventional, and flat—only involuntary recall, erupting or conjured from the depths, can convey the full quality of childhood experience, in all its innocence, wonder, and terror.

The doubling of consciousness can be confusing for Franco: the vision of Pontito, of the past, competes with the here and now, and on occasion can overwhelm it completely, so that he is disoriented, no longer knows where he is. And the doubling of consciousness has led to an odd doubling of life. Franco functions, lives, works in present-day San Francisco, but a large part of him—perhaps the larger part—is living in the past, in Pontito. And with this heightening and intensification

---

every creative artist: Kay Redfield Jamison, in *Touched with Fire*, cites Balzac, Baudelaire, Beddoes, Berlioz, Blake, Boswell, Brook, Bruckner, Bunyan, Burns, and *all* the Byrons and Brontës as manic-depressive, to name only the "B"s. It may well be that many of these attributions are correct. The danger is that we may go overboard in medicalizing our predecessors (and contemporaries), reducing their complexity to expressions of neurological or psychiatric disorder, while neglecting all the other factors that determine a life, not least the irreducible uniqueness of the individual.

of living in the past there has come a certain impoverishment and depreciation of the here and now. Franco hardly goes out, hardly travels, goes to no films or theaters; he has few recreations or interests other than his art; he used to have many friends but has lost most of them by his endless talking of Pontito. He works long hours as a cook in San Francisco's North Beach; he walks around all day, oblivious of the world, in a daze of Pontito; and all his relationships have become attenuated with his obsession—all except that with his wife, Ruth, and this was based largely on her sharing his obsession. Thus it was she who opened a gallery in North Beach and named it the Pontito Gallery, she who obtained "Pontito" license plates for the car. The cost of Franco's nostalgia and art, then, has been his reduction to a sort of half existence in the present.

The psychoanalyst Ernest Schachtel, speaking of Proust, saw him as "ready to renounce all that people usually consider an active life, to renounce activity, enjoyment of the present moment, concern with the future, friendship, social intercourse" in his hunt for the "remembrance of things past." The sort of memories for which Proust sought, and for which Franco seeks, are elusive, shy, nocturnal; they cannot compete with the full light, the bustle, of daily life—thus they must be invoked, conjured up, like dreams, in quiet and darkness, in a cork-lined room, or a mental state akin to trance or reverie.

And yet it would be reductive, absurd, to suppose that temporal lobe epilepsy, seizures of "reminiscence," even if they do constitute the final trigger of Franco's visions, could be the only determinants of his reminiscence and art. The character of the man—his attachment to his mother, his tendency toward idealization and nostalgia; the actual history of his life, including the sudden loss of his childhood paradise and of his father; and, not least, the desire to be known, to achieve, to represent a whole culture—all this, surely, is equally important. What seemed to have occurred, by a singular fortuity, was the co-occurrence, the concurrence, of an acute need and a physiological state. For if his sense of exile and loss and nostalgia demanded a sort of world, a substitute for the real world he had lost, his experiential seizures now supplied what he needed, an endless supply of images from the past—or rather, an almost infinitely detailed, three-dimensional "model" of Pontito, an entire theater or simulacrum he could mentally walk about and explore, capturing new aspects, new views, wherever he looked; this, clearly, depended equally on his prodigious, preexistent powers of memory and imagery.

As I put the events of 1965 together, I was reminded of the epileptic reminiscence that had "attacked" (but so deeply served) my patient Mrs. O'C.—which provided her, while it lasted, with long-forgotten memories of her past, memories of a most precious and significant kind. But in the case of Mrs. O'C., the epileptic reminiscence tailed off in a few weeks, closing this strange, physiologically opened door to the past and leaving her, for better or worse, "normal" once again. For Franco, however, the reminiscence was not to cease, but, if anything, swelled in intensity and volume, so that he was never, after this point, really "normal" again. Such a taking over, a possession or dispossession, occurs in a number of people with temporal lobe epilepsy—sometimes greatly heightening (but more often disrupting or destroying) their lives. In Franco's case—and here again was a singular fortuity—there was the never-before-realized power to paint his visions, to convey a child's vision with the powers of maturity, and to make of his pathology, his nostalgia, an art.

One of Franco's older sisters, Antonietta, now in Holland, remembers when the family returned to the house in Pontito after the Germans had occupied it, and found things defaced and changed. Franco's mother was deeply upset, and so was Franco. This ten-year-old fatherless child said to his mother, "I shall make Pontito again for you, I shall create it again for you." And when he did his first painting—of the house where he was born—he sent it to her; in some sense he was redeeming his promise to reconstruct Pontito for her.

Franco's mother was always seen by him, and by others, as a figure of peculiar power. "She have the power to cure the children—she taught the secret to my sister Caterina," Franco told me. "She also have the power to hurt the body by looking." Such magical thinking was common in Pontito. Franco was always very close to his mother, her favorite, and became much more so with the death of his father, when they seem to have reentered a sort of pre-Oedipal, almost symbiotic intimacy and closeness. Franco sent copies of all his paintings to her, and when she died, in 1972, he was devastated. With this, he said, "I stopped completely painting." He felt it was the end of him, of his life, of his art. He did not paint for nine months. Then as he emerged, there came an urgent need to find another woman, to marry, and now he met his future wife, a young Irish-American artist. "When I met Ruth, I wanted to go back to Italy. Ruth, she pull me back. I said 'No more reason to paint now.' But Ruth, she replace my mother. If not for Ruth, I never have painted no more."

Franco had a perpetual fantasy of going back to Pontito; he constantly talked about "a reunion" and "going home," and sometimes talked as if his mother were still mysteriously alive, waiting for him in their home, waiting for his return. Yet though he had many opportunities to go back, he managed to sabotage them all. "There is something preventing his going back to Pontito," Bob Miller said. "Some force, some fear—I don't know what it is." Franco was shocked when he saw photographs of Pontito in the late seventies—the loss of the fields and orchards, the overgrowth, appalled him—and he could hardly bear to look at the photographs that Susan Schwartzenberg took in 1987. None of this was his Pontito, the Pontito of his youth, the Pontito he had hallucinated and dreamed about and painted for more than twenty years.

There was an irony and a paradox here: Franco thought of Pontito constantly, saw it in fantasy, depicted it, as infinitely desirable—and yet he had a profound reluctance to return. But it is precisely such a paradox that lies at the heart of nostalgia—for nostalgia is about a fantasy that never takes place, one that maintains itself by not being fulfilled. And yet such fantasies are not just idle daydreams or fancies; they press toward some sort of fulfillment, but an indirect one—the fulfillment of art. These, at least, are the terms that D. Geahchan, the French psychoanalyst, has used. With reference in particular to the greatest of nostalgics, Proust, the psychoanalyst David Werman speaks of an "aesthetic crystallization of nostalgia"—nostalgia raised to the level of art and myth.

There is no doubt that Franco is at once the victim and the possessor of an imagery whose power is difficult for us to conceive. He is not at liberty to misremember, nor is he at liberty to stop remembering. There beats down on him, night and day, whether he likes it or not, a reminiscence of almost intolerable power and exactness. "No one . . . has felt the heat and pressure of a reality as indefatigable as that which day and night converged upon the hapless Ireneo," Borges writes in a sketch entitled "Funes the Memorious." Such an intolerably vivid reality converges upon Franco, too.

One may be born with the potential for a prodigious memory, but one is not born with a disposition to recollect; this comes only with changes and separations in life—separations from people, from places, from events and situations, especially if they have been of great significance, have been deeply hated or loved. It is, thus, discontinuities, the great discontinuities in life, that we seek to bridge, or reconcile, or integrate, by recollection and, beyond this, by myth and art. Discontinuity and nostalgia are most profound if, in growing up, we leave or lose the place where we were born and spent our childhood, if we become expatriates or exiles, if the place, or the life, we were brought up in is changed beyond recognition or destroyed. All of us, finally, are exiles from the past. But this is particularly true for Franco, who feels himself the sole survivor and rememberer of a world forever past.

Whatever Franco's personal gifts and pathologies—his memory, his gift for painting, his seizures (perhaps), his nostalgia—he is also moved, and has been moved throughout, by a feeling and motive that transcend the personal; by a cultural need to remember the past, to preserve its meaning, or give it new meaning, in a world that has forgotten it. In brief, we see in Franco's work the art of the exile. Much art—much mythology, indeed—stems from exile.[4] Exile (from the Garden, from Zion) is a central myth in the Bible, perhaps in every religion. Exile, of course—and perhaps, though hugely transformed, a sort of nostalgia—are central dynamics in Joyce's life and work. He left Dublin, never to return, as a very young man, but the image of Dublin haunts everything he wrote: first as the literal background of *Stephen Hero, Dubliners*, and *Exiles*, and then as the increasingly mythologized and universalized backdrop of *Ulysses* and *Finnegans Wake*. Joyce's memory of Dublin was prodigious and was continually amplified and complemented by meticulous research; but it was the Dublin of his youth that inspired him—he had little interest in its later development. And so, in a more modest way, it is with Franco: Pontito is the background of all his thoughts, from the most personal, quotidian recollections to allegorical visions of Pontito as the center of a cosmic struggle between the eternal forces of good and evil.

In March of 1989, I went to Pontito, to see the village for myself and to talk to some of Franco's relatives there. I found the village itself, compared with the paintings, at once extraordinarily similar and totally different. There is an almost photographic fidelity, an amazing microscopic power of reproduction, in the way Franco recollects, thirty years later, the details of Pontito. And yet, at the same time, I was struck by the differences: Pontito is much smaller than one would think from his paintings—the streets are narrower, the houses more irregular, the church tower shorter and more squat. There are many reasons for this, one of which is that Franco paints what he saw with a child's eye, and to a child everything is taller

---

[4]Exile—from the tropical paradise where he had spent his earliest years—was to haunt Gauguin throughout his adult life, until, finally, he went to Tahiti and tried there to reclaim the childhood Eden he had once known.

Franco's first paintings of Pontito,
done soon after his illness in
1965—the one at left is of the
house where he was born.

One of Pontito's many steep, angled stairways. Though very accurate, Franco's painting (below) broadens the perspective, adding elements that a photograph (left) is unable to do.

The view from Franco's window, again showing composite perspectives.

Two of Franco's apocalyptic or "science-fiction" paintings, showing Pontito "preserved for eternity in infinite space." The first shows the intimate view from his bedroom window; the second, a green-and-gold fragment of the church garden beneath a looming planet.

and more spacious. The literalness of this child's-eye vision made me wonder whether, through some legerdemain of the brain, Franco was able, or even forced, to reexperience Pontito exactly as he had experienced it as a child; whether he was given access, a convulsive access, to the child's memories within him.

Precisely such an access to the past—a past preserved unchanged in the brain's archives—was described to Wilder Penfield, so he thought, by some of his patients with temporal lobe epilepsy. These memories could be evoked, during surgery, by stimulating the affected part of the temporal lobes with an electrode; while the patients remained perfectly conscious of being in the operating room, questioned by their surgeon, they would also feel themselves transported to a time in the past, always the same time, the same scene, for any particular individual. The actual experiences evoked during such seizures varied enormously from patient to patient: one might reexperience a time of "listening to music," another "looking at the door of a dance hall," or "lying in the delivery room at birth," or "watching people enter the room with snow on their clothes." Because the reminiscence remained constant for each patient with every seizure or stimulation, Penfield speaks of them as "experiential seizures."[5] He conceives that memory forms a continuous and complete record of life experience, and that a segment of this is evoked and played convulsively, involuntarily, during the seizures. For the most part, he feels that the particular memories activated in this way lack special significance, and are merely inconsequential segments activated at random. But on occasion, he grants that such segments might be more—might be particularly prone to

---

[5] It is now clear that though there are repetitive or reiterative elements in such seizures, there are always elements of a fantastic or dreamlike kind as well. (One such patient, described at the turn of the century by Gowers, would always see "a sudden vision of London in ruins, herself the sole spectator in this scene of desolation," before having a convulsion or losing consciousness.) Penfield's findings are discussed, and submitted to a radically different interpretation by Israel Rosenfield, in *The Invention of Memory*.

activation because they are so important, so massively represented, in the brain. Was this, then, what was happening to Franco? Was he being forced to see, convulsively, frozen segments of his own past, "photographs" from his brain's archive?

The notion that past memories endure in the brain, though in a somewhat less literal, less mechanical form, is an idea that haunts psychoanalysis—and the great autobiographers, as well. Thus Freud's favorite image of the mind was as an archaeological site, filled, layer by layer, with the buried strata of the past (but one where these layers may rise into consciousness at any time). And Proust's image of life was as "a collection of moments," the memories of which are "not informed of everything that has happened since" and remain "hermetically sealed," like jars of preserves in the mind's larder.[6] (Proust is only one of the great meditators on memory—wondering about memory goes back at least to Augustine, without any resolution, finally, as to what memory "is.")

This notion of memory as a record or store is so familiar, so congenial, to us that we take it for granted and do not realize at first how problematic it is. And yet all of us have had the opposite experience, of "normal" memories, everyday memories, being anything but fixed—slipping and changing, becoming modified, whenever we think of them. No two witnesses ever tell the same story, and no story, no memory, ever remains the same. A story is repeated, gets changed with every repetition. It was experiments with such serial storytelling, and with the remembering of pictures, that convinced Frederic Bartlett, in the 1920s and 1930s, that there is no such entity

---

[6] In *Remembrance of Things Past*, Proust writes:

A great weakness, no doubt, for a person to consist entirely in a collection of moments; a great strength also; it is dependent upon memory, and our memory of a moment is not informed of everything that has happened since; this moment which it has registered endures still, lives still, and with it the person whose form is outlined in it.

as "memory," but only the dynamic process of "remembering" (he is always at pains, in his great book *Remembering*, to avoid the noun and use the verb). He writes,

> Remembering is not the re-excitation of innumerable fixed, lifeless and fragmentary traces. It is an imaginative reconstruction, or construction, built out of the relation of our attitude towards a whole active mass of organized past reactions or experience, and to a little outstanding detail which commonly appears in image or in language form. It is thus hardly ever really exact, even in the most rudimentary cases of rote recapitulation, and it is not at all important that it should be so.

Bartlett's conclusion now finds the strongest support in Gerald Edelman's neuroscientific work, his view of the brain as a ubiquitously active system where a constant shifting is in process, and everything is continually updated and recorrelated. There is nothing cameralike, nothing mechanical, in Edelman's view of the mind: every perception is a creation, every memory a re-creation—all remembering is relating, generalizing, recategorizing. In such a view there cannot be any fixed memories, any "pure" view of the past uncolored by the present. For Edelman, as for Bartlett, there are always dynamic processes at work, and remembering is always reconstruction, not reproduction.

And yet one wonders whether there are not extraordinary forms, or pathological forms, of memory where this does not apply. What, for example, of the seemingly permanent and totally replicable memories of Luria's "Mnemonist," so akin to the fixed and rigid "artificial memories" of the past? What of the highly accurate, archival memories found in oral cultures, where entire tribal histories, mythologies, epic poems, are transmitted faithfully through a dozen generations? What of the capacity of "idiot savants" to remember books, music, pictures, verbatim, and to reproduce them, virtually unchanged, years later? What of traumatic memories that seem to replay themselves, unbearably, without changing a single detail— Freud's "repetition-compulsion"—for years or decades after the trauma? What of neurotic or hysterical memories or fantasies, which also seem immune to time? In all of these, seemingly, there are immense powers of reproduction at work, but very much less in the way of reconstruction—as with Franco's memories. One feels that there is some element of fixation or fossilization or petrification at work, as if they are cut off from the normal processes of recategorization and revision.[7]

It may be that we need to call upon both sorts of concept— memory as dynamic, as constantly revised and represented, but also as images, still present in their original form, though written over and over again by subsequent experience, like palimpsests. In this sense, with Franco, however sharp and fixed the original, there is always some reconstruction in his work as well, particularly in the most personal pictures, such as the view from his bedroom window. Here Franco brings into an intensely personal and aesthetic unity a range of buildings that cannot be seen (or photographed) all at once, but that he has observed, lovingly, at different times. He has constructed an ideal view, which has the truth of art and transcends factuality. Whatever photographic or eidetic power Franco brings to it, such a painting always has a subjectivity, an intensely personal cast, as well. Schachtel, speaking as a

---

[7] Memory can take many forms—all, in their different ways, invaluable culturally—and we should only speak of "pathology" if these become extreme. Some people have remarkable perceptual memories, for example; they seem to take in automatically and to recollect without the least difficulty all the rich details of a summer holiday, the scores of people met, the way they dressed, their talk—the thousand incidents that make up a day on the beach. Others retain no memories (and perhaps lay down no memories) of such matters, but have huge conceptual memories, in which vast amounts of thought and information are retained, in highly abstract, logically ordered form. The mind of the novelist, the representational painter, perhaps tends to the former; the mind of the scientist, the scholar, perhaps to the latter (and, of course, one may have both sorts of memory, or varying combinations). Pure perceptual memory, with little or no conceptual disposition or capacity, may be characteristic of some autistic savants.

psychoanalyst, discusses this in relation to childhood memories:

> Memory as a function of the living personality can be understood only as a capacity for the organization and reconstruction of past experiences and impressions in the service of present needs, fears, and interests. . . . Just as there is no such thing as impersonal perception and impersonal experience, there is also no impersonal memory.

Kierkegaard goes still further, in the opening of *Stages on Life's Way*:

> Memory is merely a minimal condition. By means of memory the experience presents itself to receive the consecration of recollection. . . . For recollection is ideality . . . it involves effort and responsibility, which the indifferent act of memory does not involve. . . . Hence it is an art to recollect.

Franco's Pontito is minutely accurate, in the tiniest details, and yet it is also serene and idyllic. There is a great stillness in it, a sense of peace, not least because his Pontito is depopulated, its buildings and streets are empty; the bustling, transitory people have been removed. There is something of a desolate, a postnuclear, quality. But there is also a deeper, more spiritual stillness. One cannot help feeling that something is strange here, that what is being recalled is not the actuality of childhood, as with Proust, but a denying and transfiguring vision of childhood, with the place, Pontito, taking the place of the people—the parents, the living people— who must have been so important to the child.[8] Franco is not

---

[8] In a late paper, "Constructions in Analysis," Freud speaks of the fact that patients' memories of certain highly significant events may show a strange conjunction of excessive sharpness and detail in some respects, and total deletion in others, with crucial elements (especially human ones) missing. Thus patients may recollect "with abnormal sharpness" the rooms in which an event of great importance happened, or the furniture—but not the event itself. He sees this as the result of a conflict and compromise in the unconscious, whereby important

unaware of this and will in some moods talk of the reality of childhood as he knew it—its complexities, its conflicts, its griefs, and its pains. But all this is edited out in his art, where a paradisiacal simplicity prevails. One finds the belief in a happy childhood "even in people who have undergone cruel experiences as children," Schachtel writes. "The myth of happy childhood takes the place of the lost memory of the actual . . . experience."

And yet, we cannot reduce Franco's vision to mere fantasy or obsession. There is not just a neurotic deletion in his Pontito paintings, but an imaginative bringing-out, an intensification. Eva Brann, the philosopher, likes to call memory "the storehouse of the imagination," and (like Edelman) to see memories as imaginative, as creative, from the start:[9]

> Imaginative memory not only stores for us the passing moments of perception; it also transfigures, distances, vivifies, defangs—reshapes formed impressions, turns oppressive immediacies into wide vistas . . . loosens the rigid grip of an acute desire and transforms it into a fertile design.

And it is at this point that Franco's personal, nostalgic feelings become cultural, transcendent ones. Pontito, he feels, is special in God's eyes and must be preserved from destruction

---

memory traces are brought into consciousness, but displayed onto adjacent objects of minor significance. He stresses that such reminiscences often emerge in dreams (and thereafter daydreams), as soon as the charged subject is forced upon the mind.

[9] T. J. Murray cites a similar observation made by the painter Robert Pope, who stresses also the time which must elapse between the original experience and its re-creation—a time which, for him, averaged five years, but which, for Franco, was a quarter century or more:

> During this gestation period [writes Pope], the creative faculties act as a filter where personal opaque and chaotic data is made public, transparent and ordered. This is a process of mythologizing. Myth and dream are similar: the difference is that dreams have private, personal meaning while myths have public meanings.

and corruption. It is special, too, in embodying a precious culture—a mode of building, a mode of living, that has almost vanished from the earth. He sees his mission as one of preservation: to preserve Pontito exactly as it was, above and beyond all vicissitudes and contingencies. That this is a central dynamic, or the central dynamic, is shown by a series of remarkable apocalyptic or "science-fiction" paintings, which he seems to do in periods of mental stress or distress. In these, the earth is menaced by another planet or a comet, by imminent or actual destruction, but Pontito survives: Franco shows the old church, or a garden, all green and gold, radiant, transfigured, in a beam of sunlight, miraculously surviving the all-encompassing destruction. (In another allegorical picture, he put a satellite dish on the church: a dish aimed at the stars—and at God.) These apocalyptic paintings have titles like *Pontito Preserved for Eternity in Infinite Space.*

Franco gets up early each morning and knows what he has to do. He has his task, his mission: to recollect—to consecrate the memory of Pontito. His visions, when they come, are full of emotion and excitement—no less so than they were when they first came to him, twenty-five years ago. And the activity of painting—of walking again in recollection through the so-loved paths and streets, and being able to articulate this, in so masterly a fashion, with such richness and detail—gives him a sense of identity and accomplishment by giving his visions a controlled and artistic form.

"I don't feel that I deserve the credit for these paintings," Franco wrote me in a letter. "I did them for Pontito. . . . I want the whole world to know how fantastic and beautiful it is. In this way maybe it won't die, although it is dying. Maybe my paintings will at least keep its memory alive."

Up to early 1989, I had seen Franco and visited him at his house in San Francisco several times; I had spoken with his friends there; I had met two of his sisters in Holland; and, above all, I had visited Pontito, which excited and teased Franco, for he was thinking now, more than at any time in the past twenty years, of visiting Pontito himself. His life had had, until now, a strange sort of stability, with living, eating, functioning—somewhat absentmindedly—in the present, but with his mind and art constantly fixed on the past. In this he had been greatly aided by Ruth, who, though herself an artist, had identified herself in the deepest way with Franco's Pontitan relationship and art and did all she could to take care of the mundane necessities of life and to give him the protection and insulation he needed to dwell and work uninterrupted in his nostalgic art. But in 1987, tragically, she became sick, and, after a painful fight with cancer, she died, just three months before Franco's Exploratorium exhibit. This was his first big show, and, along with his wife's death, it stirred feelings that he could no longer go on as he had in the past—something new must happen, new decisions must be made. He sounded these themes in a letter he wrote me a month later:

> Very shortly I may be moving. Probably to San Francisco, but maybe back to Italy for good. . . . My situation since my wife died has been difficult for me. I'm not sure what I should do. . . . I must sell my house, look for a new place and job in San Francisco, or in the future go back to Pontito. So that will be the end of the Pontito memory—but not the end of my life! I'll start a new memory.

I was struck by the way in which he equated a return to Pontito with the end of his memory, his identity, the end of his singular reminiscence and art. One saw now why he had sabotaged all previous opportunities to go there. Could the fairy tale, the myth, survive reality?

In March of 1989, I spoke of Franco and his art at a conference in Florence. Invitations started to pour in on Franco—to give interviews, to send slides, to allow an exhibit, and, above all, to return to Pontito. Pescia, the nearest big town to Pontito, organized an exhibit of his paintings, to be held in September 1990. His long-standing inner conflict was magnified by this outside notice; a state of excitement and ambiva-

lence and agitation grew. Finally, that summer, he decided
to go.

He had envisaged walking from Pescia—walking up the
winding mountain road to Pontito, carrying on his back a
wooden cross he had made, which he would place in the old
church at Pontito. He would be alone, utterly alone, in this
consecrated walk. He would stop at a spring, an ancient
spring of fresh water, just outside Pontito, and put his face in
the gushing waters. Perhaps after drinking the waters, he
thought, he would lie down and die. Or perhaps, purified, born
again, he would reenter Pontito. No one would recognize him,
the grizzled stranger from afar, until an old dog—the old dog
he had known as a child, now so old it could scarcely move
(the dog, indeed, would have to be the same age as Franco
himself)—until his old dog, recognizing him, would feebly lick
him and then, its waiting over, would wag its tail and die. It
was singular to hear this elaborate fantasy from Franco, this
fantasy with elements of Sophocles and Homer no less than
the New Testament, for he had never read, never heard of,
Sophocles or Homer.

In the event, his return was nothing like this.

He had phoned me in a panic the evening before his flight.
Innumerable thoughts and desires and fears were colliding inside him: Should he go, or shouldn't he? He kept changing his
mind. Since his art was based on fantasy and nostalgia, on a
memory uncontaminated by updating, he was terrified that he
would lose it if he returned to Pontito. I listened carefully,
like an analyst, offering no suggestions. "*You* have to decide,"
I said, finally. He took the red-eye flight later that evening.

He had hoped that, first, he might meet the Pope and have
his cross blessed before he walked with it to Pontito. But the
Pope was away, in Africa. Nor was the Via Dolorosa walk to
Pontito possible. The mayor of Pescia and other officials were
at this moment, he was told, awaiting him in Pontito, and he
was whisked off there, at high speed, in a car.

The ceremony over, Franco took off by himself, going to his

boyhood home. His first impression: "Oh my God, it was so small I had to crouch to look through my window. I see changes outside—but to me is no change." As he walked around the town, it seemed uncannily quiet, deserted, "like everybody is left, like the town is mine." He savored, for a little, this sense that it was his, and then got a sense of grievous loss: "I missed the chickens, the donkey shoes. Like a dream. Everybody left. You used to hear a lot of noise—the children coming up, the women, the donkey shoes. All gone." No one greeted him, no one recognized him, no one was to be seen on the streets during this first walk. He saw no curtains in the windows and no laundry hanging, heard no sounds of life coming from the empty, shuttered houses. He encountered only half-feral cats slinking in the alleys. The feeling grew on him that Pontito was indeed dead, and he himself a revenant returning to a ghost town.

He wandered out beyond the houses, into the areas that used to be lush with well-tended fields and orchards. Everywhere the ground was cracked and dry; there was neglect, and a huge overgrowth of parasites and weeds. Now, it seemed to him, not only Pontito but the whole enterprise of civilization was in ruins. He thought of his own apocalyptic visions: "Someday it will be polluted, overgrown. There will be nuclear war. So I will put it in Space, to be preserved for Eternity."

But then, as the sun rose, the sheer beauty of the scene made him catch his breath: "I can't believe it, it's so beautiful." There, rising up tier by tier on the mountain, was Pontito, his Pontito, all green and gold, the church tower at its crest glinting now in the early morning sun—*his* church, completely unchanged. "I went up in the tower. I touched the stones. Its age to me is like a thousand years. All different colors—the copper, the green." Touching the stones, stroking them, caressing them, Franco grounded himself, started to feel again that Pontito was real. Stones have always played a crucial role in his paintings; they are portrayed with the utmost accuracy—every shade, every color, every convexity or crack, lovingly dwelt on and delineated. There is an extraordinary tactile or kinesthetic quality in Franco's stones. Now, as he touched them, the actuality of "coming home" started to return, and for the first time during his visit Franco started to rejoice. The stones, at least, had not changed. Nor had the church, or the buildings, or the streets. Their feel, at least, was still what it had been. And now the villagers, many of them relatives, came out of their houses, excitedly greeting him, bombarding him with questions. Everyone was proud of him: "We've seen your paintings, we've been hearing about you—you're coming back to us now?" And now he started to feel like the prodigal son. This, he said later, was a high point of his trip: "As a young child in Pontito, I thought, One day I'm going to grow up. Do something, be somebody, for my *madre*. Show the people in Pontito. After my father died, I had no shoes, all broke. I used to feel shame. We were despised."

His childhood fantasy was coming true: Franco had done something, was somebody, and now people—not just people in America or Italy, but his own people, the Pontitans—loved and admired him. A tender feeling for the people—"my people"—seized him. They could not remember the past as he did—their memories lacked the power of his or had been updated, effacing the past. This was evident whenever he spoke to them. He, then, would be their archive, their memory: "I bring back the memory to these people." And he later said to the mayor, "I'm going to build a gallery, a little museum, something to bring the people back to the town."

On the surface, returning to Pontito was not as intense an experience as he had expected—there had been no mystical revelations, no ecstasies on the heights—but neither had he dropped dead from poisoned waters or had a heart attack, as he had also more than half expected. It was when he left that he really felt the impact.

Back in San Francisco, he found himself in a crisis. First, there was an overwhelming sensory confusion: he seemed to see two pictures of Pontito—two "newsreels," as he put it— running simultaneously in his head, with the more recent, the new, tending to blot out the old. He could do nothing to stop this perceptual conflict, and when he tried to paint Pontito he found that he no longer knew what to do: "I get confused, I see these two pictures at once," he told me. "I thought I would paint Pontito as it was, but I 'see' it as it is now. I thought I would go crazy. What could I do? Maybe I could never paint Pontito again. I got scared. My God, now—start all over again? . . . It took me ten days to come back."

It took ten days for the hallucinatorily vivid pictures of the new Pontito to die down, to stop competing with the old Pontito; ten days for the merely sensory conflict to resolve; and, as for his emotions, they were so confused he hardly dared think about them. At this point, almost desperate, he said, "I wish I never went back. I work best with my fantasy. I can't work now." It was a month before he started to draw Pontito again. These new drawings and paintings, just a few inches square, took on an unusually tender and intimate quality: corners, nooks where a boy might sit, nooks where he had sat and dreamt as a child. These little scenes, though they did not contain human figures, had an intensely human feel, as if their occupants had just left or were just about to arrive—very different from the idealized yet deserted scenes he had usually painted.

Thinking over the experience, Franco felt that it had been both enjoyable and exhausting, but compromised, at a deeper level, in his three weeks there, because he had had no time to himself—he had been followed and interviewed every day in Pontito and had had no time to sketch or think. He felt a need to go back a second time, to confront the deeper issues, to spend time alone in Pontito.

In March 1991, there was a second exhibit of Franco's paintings in Italy—this one in the Palazzo Medici-Riccardi in Florence—and I accompanied Franco to the exhibit. He was abashed by the splendid surroundings, by seeing his paintings in vast, palatial rooms. "I feel like an intruder," he said. "They do not belong." He and his paintings, he feels, are rooted in the hilly Tuscan countryside; he feels uncomfortable in the cosmopolitan grandeur of Florence.

The next morning, Franco and I are off to Pontito; for the first time, we will see his town together. We pass the Duomo and the Baptistry in the center of Florence, pass the old children's hospital, the Innocenti, driving through the miraculously preserved old city, unspoiled and deserted now, near dawn on a Sunday. Franco, beside me, is rapt, absorbed in his thoughts.

We pass the road for Pistoia and head toward Montecatini, the slopes on either side of us dotted with old hill towns. "There is at the back of every artist's mind something like a pattern or a type of architecture," G. K. Chesterton wrote. "It is a thing like the landscape of his dreams; the sort of world he would wish to make or in which he would wish to wander; the strange flora and fauna of his own secret planet." For Auden, this landscape was limestone and lead mines; for Franco, it is this old, gnarled, unchanging Tuscan landscape.

A sign warning motorists of snow prompts me to ask Franco whether there was ever snow in Pontito, or whether he had ever painted a snow-whitened Pontito. Yes, there was snow, he says, and he once started a snowscape, but almost all his paintings are of Pontito *in primavera*, in spring.

As we reach Pescia, at the bottom of the mountain, below Pontito, Franco recognizes people and places: the shop where he used to buy paints forty years ago; a subterranean bar. Little has changed in this slow-paced town. He recognizes the mailman from the 1940s: they throw their arms around each other in the street. Everyone is welcoming; there are smiles everywhere for the prodigal son come home once more. We move on to the city hall, where Franco was given the honors of the city during his first visit. A prophet is honored now in his own land. This pleases him, this local fame; he belongs here, as he does not belong in Florence.

From Pescia the road is narrow and steep. We wind up in second gear after nearly ditching ourselves on the first turn, past Pietrabuona, a town named after its fine stone, with its church and oldest buildings perched on its highest hill. We pass its terraced hills, softly lit, with gnarled olive trees and vines upon them; these terraces are ancient, dating from Etruscan times. We wind around past many small villages— Castelvecchio, Stiappa, San Quirico. Finally, we round another bend in the road and catch our first sight of Pontito. "My God, look at it!" Franco exclaims, sotto voce. "Jesus Christ! I can see my home. No, I can't. . . . This overgrowth is bad, parasites everywhere. Used to be cherry, pear, fruit trees. Chestnuts, grains, corn, lentils." He tells me how, as a lanky, long-legged youth, he used to stride from one village to another. As we approach Pontito, Franco's eyes grow moist. He stares intensely and murmurs to himself as he takes everything in. "This is the bridge, the stream where we did the washing. Down the path, here, the women would walk with baskets on their heads."

We stop the car, and Franco leaps out, seeing and remembering more details all the time. And, along with this pure topographic memory, there is also a cultural one. He describes how the villagers would take hemp and immerse it in the stream for a year, anchored by rocks, and then take it out to be dried and woven into fabric for sheets and towels, and for sacks for chestnuts: a whole local industry, a tradition, now nearly forgotten, except by Franco. Suddenly, indignant at new growth obscuring the path, he tears it out in giant armfuls. Angry at some new building, he tells me in detail exactly how it used to be: "There was a big rock there, the water ran here." There is no doubt that every stone, every inch, is engraved in his memory.

"*Come sta?*" Climbing up the steep cobbled street, Franco greets a stout middle-aged man in a green coat. ("His father gave us candies.") Franco has a bardic memory, but the trivial and the momentous, the personal and the mythic, are indiscriminately mixed. He stops at the house where his mother was born.

"Sabatoni!"

"Franco!" An old man emerges. ("It's my uncle.") "You've been in America. What brings you back? I heard there was a show in Florence." The old man mentions the drying of chestnuts. He forgets the details, but Franco does not. The old man points out that the four houses next to his, so full of life once, are now empty. "When I am dead, it will be empty here, too."

We visit Franco's sister Caterina. She and her husband have retired to Pontito, and Franco is distressed to see her looking older than he remembers. Caterina feeds us a magnificent Tuscan lunch—cheese, bread, olives, wine, tomatoes preserved from her garden—and then Franco takes me off to look at the church. It is a beautiful spot, atop the hill and overlooking the rest of the village. In the cemetery, Franco points out the graves of his mother, his father, this relative and that. "There are more people in the graveyard than in the town," he says softly. Franco plans to stay in Pontito for three more weeks, to do some quiet sketching. He says, "I'm going to put my roots back here." But, as I leave, my final image is of Franco standing by himself in the cemetery, gazing over the depopulated town, alone.

Franco's three weeks in Pontito seemed to recharge him; at least, he has been incessantly active since his return. His garage-studio is crackling with life. There are pictures everywhere, old and new—the new ones based on sketches he did in March, and the old ones, started in 1987 but left unfinished with Ruth's death, now being completed in a burst of new decision and energy.

Seeing Franco once again at work, his renewed fury of recollective and creative energy, raises anew all the questions one has about his singular enterprise, the *meaning* of Pontito for him. His "new" paintings are not really new—he may add the new here and there (a fence, a gate, a new tree perhaps), but they remain essentially the same. His project, in a fundamental sense, remains unchanged. When I visited Franco last summer, I saw a pair of sneakers hanging from the rafters of his garage, with an elaborately calligraphed notice tied to them, saying, in Italian, "With these shoes, after 34 years, I first set foot in what had been the Promised land." Now that he had set foot in it, it had lost some of its glamour, its promise. "Sometimes I wish I never went back," he said as he saw me looking at the shoes. "Fantasy, memory, that is the most beautiful." And then he added, musingly, "Art is like dreaming."

Seeing the current reality of Pontito was very disturbing to Franco, although he was able to recover from the derailment it caused. But it heightened his sense that the Pontito of here and now is a threat to his own vision and showed him that he must ration any further exposure to it. There have been many subsequent invitations, but he has not returned, even for an exhibit of his own work in the streets of Pontito. Other artists now are flocking to Pontito, but for them it is just another charming Tuscan hill town. Franco, fleeing all this, has returned to his garage, returned to the project that has consumed him for twenty-nine years. It is a project that has no end, can never be brought to a conclusion or completion, and he paints now, one sometimes feels, in a sort of frenzy, barely finishing one canvas before moving on to another. He is experimenting with other forms of representation as well: cardboard models of Pontito, which he fashions with his long agile fingers, and videotapes of his paintings (accompanied by music) to simulate a walk through the town. He is fascinated by the idea of computer simulations of Pontito and the thought that one might don helmet and gloves—and not only see, but *touch* its virtual reality, too.

When I met him originally, Franco was billed as "A Memory Artist," implying his affinity to Proust, "the poet of memory." At first I thought there was indeed a similarity—both men, both artists, withdrawing themselves from the world, in order to recapture the lost world of childhood. But now one sees, increasingly with each year, how totally Franco's project differs from Proust's. Proust, too, was haunted by the lost, the forgotten past, and his quest was to find if the door to it could be opened. As he succeeded in this, partly through the grace of "involuntary memories," partly through vast intellectual labor, his work could reach its completion and conclusion (a completion at once psychological and artistic).

But this is not possible for Franco, who instead of achieving a penetration into the inwardness, the "meaning," of Pontito, makes a vast, even infinite enumeration of all its outward aspects—its buildings, its streets, its stones, its topography—as if these could in some way compensate for the human void within. He half knows this, yet does not know it, and in any case has no choice. He has no time for, no taste for, no power of introspection and may suspect, indeed, that it would be fatal to his art.

Franco feels he has twenty, thirty years of work still ahead of him, for the thousand-odd paintings he has done since 1970 convey only a part of the reality he seeks to portray. He has to have paintings, or simulations, of every detail, from every viewpoint—from the village in the distance, as one drives up to it from Pistoia, to the finest details of the lichened stones in the church. He envisions the building of a museum overlooking the town, which will house a vast archive of Pontito, his Pontito—the thousands of paintings he has made and the thousands more he still intends to make. It will be the culmination of his life's work, and the redemption of his promise to his mother: "I shall create it again for you."

Prodigies

The Fayetteville *Observer* of May 19, 1862, contained an unusual letter from its correspondent Long Grabs, stationed in Camp Mangum:

> The blind negro Tom has been performing here to a crowded house. He is certainly a wonder.... He resembles any ordinary negro boy 13 years old and is perfectly blind and an idiot in everything but music, language, imitation, and perhaps memory. He has never been instructed in music or educated in any way. He learned to play the piano from hearing others, learns airs and tunes from hearing them sung, and can play any piece on first trial as well as the most accomplished performer.... One of his most remarkable feats was the performance of three pieces of music at once. He played Fisher's Hornpipe with one hand and Yankee Doodle with the other and sang Dixie all at once. He also played a piece with his back to the piano and his hands inverted. He performs many pieces of his own conception—one, his "Battle of Manassas," may be called picturesque and sublime, a true conception of unaided, blind musical genius.... This poor blind boy is cursed with but little of human nature; he seems to be an unconscious agent acting as he is acted on, and his mind a vacant receptacle where Nature stores her jewels to recall them at her pleasure.

We learn more of Blind Tom from Edouard Séguin, the French physician whose 1866 book, *Idiocy and Its Treatment by the Psychological Method*, contained many penetrating descriptions of individuals later to be termed "idiots savants"; and from an intellectual descendant of Séguin, Darold Treffert, whose book *Extraordinary People: Understanding "Idiot Savants"* was published in 1989. Born nearly blind, the fourteenth child of a slave, sold to a Colonel Bethune, Tom was, from infancy, Treffert writes, "fascinated by sounds of all sorts—rain on the roof, the grating of corn in the sheller, but most of all music—Tom would listen intensely to the colonel's daughters practicing their sonatas and minuets on the piano."

"Till five or six years old," Séguin writes, "he could not speak, scarce walk, and gave no other sign of intelligence than this everlasting thirst for music. At four years already, if taken out from the corner where he lay dejected, and seated at the piano, he would play beautiful tunes; his little hands having already taken possession of the keys, and his wonderful ear of any combination of notes they had once heard." At the age of six, Tom started to improvise on his own account. Word of the "blind genius" spread, and at seven Tom gave his first concert—and went on to earn a hundred thousand dollars in his eighth year. At eleven, he played before President Buchanan at the White House. A panel of musicians, who thought that he had tricked the president, tested his memory the following day, playing two entirely new compositions to him, thirteen and twenty pages in length—he reproduced them perfectly and without the least apparent effort.

Séguin, describing Tom listening to a new piece, adds further tantalizing details in regard to his expressions, postures, and movements:

> [He] shows his satisfaction by his countenance, a laughing, stooping, with various rubbings of the hand, alternating with an increase of the sideway swinging of his body, and some uncouth smiles. As soon as the new tune begins, Tom takes some ludicrous posture [with one leg outstretched, while he slowly pirouettes on the other] . . .

long gyrations ... ornamented with spasmodic movements of the hands.

Although Tom was usually called an idiot or imbecile, such posturing and stereotypies are more characteristic of autism—but autism was only identified in the 1940s and was not a term, or even a concept, in the 1860s.

Autism, clearly, is a condition that has always existed, affecting occasional individuals in every period and culture. It has always attracted in the popular mind an amazed, fearful, or bewildered attention (and perhaps engendered mythical or archetypal figures—the alien, the changeling, the child bewitched). It was medically described, almost simultaneously, in the 1940s, by Leo Kanner in Baltimore and Hans Asperger in Vienna. Both of them, independently, named it "autism."

Kanner's and Asperger's accounts were in many ways strikingly (at times uncannily) similar—a nice example of historical synchronicity. Both emphasized "aloneness," mental aloneness, as the cardinal feature of autism; this, indeed, was why they called it autism. In Kanner's words, this aloneness "whenever possible, disregards, ignores, shuts out anything that comes to the child from the outside." This lack of contact, he felt, was only in regard to people; objects, by contrast, might be normally enjoyed. The other defining feature of autism, for Kanner, was "an obsessive insistence on sameness," in the form of repetitive, stereotyped movements and noises, or stereotypies, most simply; then in the adoption of elaborate rituals and routines; finally, in the appearance of strange, narrow preoccupations—highly focused, intense fascinations and fixations. The appearance of such fascinations, and the adoption of such rituals, often before the age of five, were not to be seen, Kanner and Asperger thought, in any other condition. Asperger brought out other striking features, stressing,

> they do not make eye contact ... they seem to take in things with short, peripheral glances ... [there is] a poverty of facial expressions and gestures ... the use of lan-

guage appears abnormal, unnatural . . . the children follow their own impulses, regardless of the demands of the environment.

Singular talents, usually emerging at a very early age and developing with startling speed, appear in about 10 percent of the autistic (and in a smaller number of the retarded—though many savants are both autistic and retarded). A century before Blind Tom there was Gottfried Mind, a "cretinous imbecile," born in Berne in 1768, who showed from an early age a striking talent for drawing. He had, according to A. F. Tredgold's classic 1908 *Text-Book of Mental Deficiency*, "such a marvellous faculty for drawing pictures of cats that he was known as 'The Cats' Raphael,' " but he also made drawings and water-color sketches of deer, rabbits, bears, and groups of children. He soon acquired fame throughout Europe, and one of his pictures was purchased by George IV.

Prodigious calculators attracted attention in the eighteenth century—Jedediah Buxton, a simpleminded laborer, had perhaps the most tenacious memory of these. When asked what would be the cost of shoeing a horse with a hundred and forty nails if the price was one farthing for the first nail, then doubled for each remaining nail, he arrived at the (nearly correct) figure of 725,958,096,074,907,868,531,656,993,638,851,106 pounds, 2 shillings, and 8 pence. When he was then asked to square this number (that is, 2^{139} squared), he came up with (after two and a half months) a seventy-eight digit answer. Though some of Buxton's calculations took weeks or months, he was able to work, to hold conversations, to live his life normally, while doing them. The prodigious calculations proceeded almost automatically, only throwing their results into consciousness when completed.

Child prodigies, of course, are not necessarily retarded or autistic—there have been itinerant calculators of normal intelligence as well. One such was George Parker Bidder, who as a child and youth gave exhibitions in England and Scotland. He could mentally determine the logarithm of any number to seven or eight places and, apparently intuitively, could divine the factors for any large number. Bidder retained his powers throughout life (and indeed made great use of them in his profession as an engineer) and often tried to delineate the procedures by which he calculated. In this, however, he was unsuccessful; he could only say of his results that "they seem to rise with the rapidity of lightning" in his mind, but that their actual operations were largely inaccessible to him.[1] His son, also intellectually gifted, was a natural calculator as well, though not as prodigious.

Besides these major domains of savant expertise, some savants have astonishing verbal powers—the last thing one might expect in intellectually defective individuals. Thus there are savants who are able, by the age of two, to read books and newspapers with the utmost facility but without the least comprehension (their expertise, their decoding, is wholly phonological and syntactic, without any sense of meaning).

Almost all savants have prodigious powers of memory. Dr. J. Langdon Down, one of the greatest observers in this realm, who coined the term "idiot savant" in 1887, remarked that "extraordinary memory was often associated with very great defect of reasoning power." He describes giving one of his patients Gibbon's *Decline and Fall of the Roman Empire*. The patient had read the entire book and in a single reading imprinted it in memory. But he had skipped a line on one

---

[1] Later, Bidder described some of the techniques and algorithms which he found himself using; though their discovery in the first place, as well as their use, seemed to be unconscious. In our own time, A. C. Aitken, a great mathematician and calculator, observes:

I have noticed at times that the mind has anticipated the will; I have had an answer before I even wished to do the calculation; I have checked it, and am always surprised that it is correct. This, I suppose (but the terminology may not be right), is the subconscious in action; I think it can be in action at several levels; and I believe that each of these levels has its own velocity, different from that of our ordinary waking time, in which our processes of thought are rather tardy. (This is cited by Steven B. Smith in "Calculating Prodigies.")

page, an error at once detected and corrected. "Ever after," Down tells us, "when reciting from memory the stately periods of Gibbon, he would, on coming to the third page, skip the line and go back and correct the error with as much regularity as if it had been part of the regular text." Martin A., a savant I wrote about in "A Walking Grove," could recall the entire nine volumes of Grove's 1954 *Dictionary of Music and Musicians*. This had been read to him by his father, and the text would be "replayed" in his father's voice.

There is a large variety of minor savant skills, frequently described by physicians like Down and Tredgold, who consulted at institutions for the "mentally defective." Tredgold describes J. H. Pullen, "the Genius of Earlswood Asylum," who for more than fifty years made extremely intricate models of ships and machines, as well as a very real guillotine, which almost killed one of his attendants. Tredgold writes of an otherwise retarded savant who could "get" a complex mechanism like a watch and disassemble and reassemble it swiftly, with no prior instruction. More recently, physicians have described idiot savants with extraordinary bodily skills, able to perform acrobatic maneuvers and athletic feats with the greatest facility—again, with no formal training. (In the 1960s, I saw, on a back ward, such a savant myself—he had been described to me as "an idiot Nijinsky.")[2]

While early medical observers sometimes conceived of savant skills as the hypertrophy of a single mental faculty, there was little sense that savant talents were of much more than anecdotal interest. An exception here was the eccentric psychologist F. W. H. Myers, who, in his great turn-of-the-century book, *Human Personality*, tried to analyze the processes by

---

[2] Tredgold writes of savants with various sensory powers and skills, of olfactory savants—and of a tactile savant, too:

Dr. J. Langdon Down told me of a boy at Normansfield whose sense of touch was so delicate and his fingers so deft that he could take a sheet of the Graphic and gradually split it into two perfect sheets, as one would peel a postage stamp off an envelope.

which prodigious calculators arrived at their results. He was unable to do so, any more than could the calculators themselves, but he believed that a process of "subliminal" mentation or computation was involved, which threw its results into consciousness when complete. Their methods of calculation seemed to be—unlike the formal or formulary methods taught in primers and schools—idiosyncratic and personal, achieved by each calculator through an individual path. Myers was one of the first to write about unconscious or preconscious cognitive processes and foresaw that an understanding of idiots savants and their gifts could open not only into a general understanding of the nature of intelligence and talent but into that vast realm that we now call the cognitive unconscious.

In the 1940s, when autism was first delineated, it became evident that the majority of idiot savants were in fact autistic and that the incidence of savantism in the autistic—nearly 10 percent—was almost two hundred times its incidence in the retarded population, and thousands of times that of the population at large. Furthermore, it became clear that many autistic savants had multiple talents—musical, mnemonic, visual-graphic, computational, and so on.

In 1977, the psychologist Lorna Selfe published *Nadia: A Case of Extraordinary Drawing Ability in an Autistic Child.* Nadia suddenly started drawing at the age of three and a half, rendering horses, and later a variety of other subjects, in a way that psychologists considered "not possible." Her drawings, they felt, were qualitatively different from those of other children: she had a sense of space, an ability to depict appearances and shadows, a sense of perspective such as the most gifted normal child might only develop at three times her age. She constantly experimented with different angles and perspectives. Whereas normal children go through a developmental sequence from random scribbling to schematic and geometric figures to "tadpole" figures, Nadia seemed to bypass these and to move at once into highly recognizable, detailed representational drawings. The development of drawing in children, it was felt at the time, paralleled the development of conceptual powers and language skills; but Nadia, it seemed, just drew what she saw, without the usual need to "understand" or "interpret" it. She not only showed enormous graphic gifts, an unprecedented precocity, but drew in a way that attested to a wholly different mode of perception and mind.[3]

The case of Nadia—set out at monographic length and minutely documented—aroused great excitement in the neurological and psychological communities and suddenly focused a belated attention on savant talents and on the nature of talents and special abilities in general. Where, for a century or more, neurologists had confined their attention to failures and breakdowns of neural function, there was now a move in the other direction, to exploring the structure of heightened powers, of talents, and their biological basis in the brain. Here idiot savants provided unique opportunities, for they seemed to exhibit a large range of inborn talents—raw, pure expressions of the biological: much less dependent upon, or influenced by, environmental and cultural factors than the talents of "normal" people.

In June of 1987, I received a large packet from a publisher in England. It was full of drawings, drawings that delighted me greatly because they portrayed many of the landmarks I had grown up with in London: monumental buildings like St.

---

[3] Though prodigious musical abilities tend to show themselves extremely early—almost all the great composers exemplify this—"there are no prodigies in art," as Picasso said. (Picasso himself was a remarkable draftsman at ten, but could not draw horses at three, like Nadia, or cathedrals at seven.) There must be fundamental neurodevelopmental and cognitive reasons for this. Though Yani, a nonautistic Chinese girl, showed her artistic powers very early—she had done thousands of paintings by the age of six—her paintings are those of a very gifted, sensitive (and highly trained) child, arising from a normal, albeit accelerated, perceptual development, which was undoubtedly encouraged by her artist father. Her paintings are quite unlike the suddenly appearing, full-blown, "unchildlike" drawings characteristic of prodigious graphic savants like Stephen Wiltshire. There may, of course, exist in some nonautistic people a mixture of savant and normal talents (see footnote 9, page 226).

Paul's, St. Pancras Station, the Albert Hall, the Natural History Museum; and others, odd, sometimes out of the way, but dear and familiar places, like the Pagoda in Kew Gardens. They were very accurate, but not in the least mechanical—on the contrary, they were full of energy, spontaneity, oddity, life.

In the packet, I discovered a letter from the publisher: the artist, Stephen Wiltshire, was autistic and had shown savant abilities from an early age. His *London Alphabet*, a sequence of twenty-six drawings, had been done when he was ten; an amazing elevator shaft, with a vertiginous perspective, when he was eight. One drawing was an imagined scene, of St. Paul's surrounded by flames in the Great Fire of London. Stephen was not merely a savant but a prodigy. Sixty of his drawings, a mere fraction of what he had done, were to be published, the letter informed me; the author was just thirteen.

Stephen's drawings reminded me, in many ways, of drawings by my patient José—"The Autist Artist" whom I had known and written about, years before—with an extraordinary eye and gift for drawing. Though José and Stephen came from such different backgrounds, the similarity of their drawings was so uncanny as to make me wonder whether there might not be a distinctive "autistic" form of perception and art. But José, despite his fine gifts (not, perhaps, as great as Stephen's, but quite remarkable nonetheless), was wasting away in a state psychiatric hospital; Stephen had somehow been luckier.

A few weeks later, visiting family and friends in England, I mentioned Stephen and his drawings to my brother, David, a general practitioner in northwest London. "Stephen Wiltshire!" he exclaimed, very startled. "He's a patient of mine—I've known Stephen since he was three."

David told me something of Stephen's background. He was born in London in April of 1974, the second child of a West Indian transit worker and his wife. Unlike his older sister, Annette, born two years earlier, Stephen showed some delay in the motor landmarks of infant life—sitting, standing, hand control, walking—and a resistance to being held. In his second and third years, more problems appeared. He would not play with other children and tended to scream or hide in a corner if they approached. He would not make eye contact with his parents or anyone else. Sometimes he seemed deaf to people's voices, though his hearing was normal (and thunder terrified him). Perhaps most disquieting, he did not use language; he was virtually mute.

Just before Stephen's third birthday, his father was killed in a motorcycle accident. Stephen had been strongly attached to him and after his death grew much more disturbed. He started screaming, rocking, and flapping his hands and lost what little language he had. At this point the diagnosis of infantile autism had been made, and arrangements made for him to attend a special school for developmentally disabled children. Lorraine Cole, the headmistress at Queensmill, observed that Stephen was very remote when he started school at the age of four. He seemed unaware of other people and showed no interest in his surroundings. He would simply wander about aimlessly or occasionally run out of the room. As Cole writes:

> He had virtually no understanding of or interest in the use of language. Other people held no apparent meaning for him except to fulfill some immediate, unspoken need; he used them as objects. He could not tolerate frustration, nor changes in routine or environment and he responded to any of these with desperate, angry roaring. He had no idea of play, no normal sense of danger and little motivation to undertake any activity except scribbling.

She later wrote to me, "Stephen would climb onto a play-bike, pedal it furiously, then hurl himself off it, roaring with laughter, and sometimes screaming."

Yet at this point the first evidence of his visual preoccupation, and talent, appeared. He seemed fascinated by shadows, shapes, angles, and by the age of five he was fascinated by pictures, too. He would make "sudden dashes to other rooms, where he would stare intently at pictures which fascinated him," Cole writes. "He would find paper and pencil and scribble, totally absorbed for long periods."

Stephen's "scribblings" were largely of cars and occasionally of animals and people. Lorraine Cole speaks of his doing "wickedly clever caricatures" of some of the teachers. But his special interest, his fixation, which developed when he was seven, was the drawing of buildings—buildings in London he had seen on school trips or that he had seen on television or in magazines. Why he developed this sudden, special interest and preoccupation so powerful and exclusive that he now had no impulse to draw anything else is not wholly clear. Such fixations are exceedingly common in autistic people. Jessy Park, an autistic artist, is obsessed with weather anomalies and constellations in the night sky;[4] Shyoichiro Yamamura, an autistic artist in Japan, drew insects almost exclusively; and Jonny, an autistic boy described by the pioneer psychologist Mira Rothenberg, for a period drew only electric lamps, or buildings and people composed of electric lamps. Stephen, from this very early age, had been almost exclusively preoccupied with buildings—buildings, by preference, of great complexity and size—and also with aerial views and extremities of perspective. He had one other interest at the age of seven: he was fascinated by sudden calamities, and above all by earthquakes.

---

[4] Meeting a young physicist-astronomer, Ben Oppenheimer, recently, I mentioned Jessy's paintings, and showed him copies of some. He was astounded at their astronomical accuracy, and was reminded of an amateur astronomer and minister, Robert Evans, in Australia. Evans single-handedly, with a small telescope, observed the incidence of supernovae in a sample of 1017 bright (Shapley-Ames) galaxies which he observed for a period of five years (examining, Oppenheimer calculates, sixty or more galaxies each night); he went on from this to deduce a new figure for the supernova rate in such galaxies. (This work was published by van den Bergh, McClure, and Evans in *The Astrophysical Journal*.) Evans used no photographic or electronic assistance, and thus seemed able to construct and hold in his mind an absolutely precise and stable image or map of more than a thousand galaxies, as seen in the southern sky. It seems likely that his memory is either eidetic or savantlike, though there is no suggestion that he is autistic.

Whenever Stephen drew these, or saw them on television or in magazines, he grew strangely excited and overwrought—nothing else disturbed him in quite this way. One wonders whether his earthquake obsession (like the apocalyptic fantasies of some psychotics) represented a sense of his own inner instability, which in drawing he could try to master.

When Chris Marris, a young teacher, came to Queensmill in 1982, he was astonished by Stephen's drawings. Marris had been teaching disabled children for nine years, but nothing he had ever seen had prepared him for Stephen. "I was amazed by this little boy, who sat on his own in a corner of the room, drawing," he told me. "Stephen used to draw and draw and draw and draw—the school called him 'the drawer.' And they were the most unchildlike drawings, like St. Paul's and Tower Bridge and other London landmarks, in tremendous detail, when other children his age were just drawing stick figures. It was the sophistication of his drawings, their mastery of line and perspective, that amazed me—and these were all there when he was seven."

Stephen was one of a group of six in Chris's class. "He knew the names of all the others," Chris told me, "but there was no sense of interaction or friendship with them. He was such an isolated little chap." But his native gift was so great, Chris felt, that he did not need to be "taught," in the ordinary way. He had apparently worked out by himself, or had an innate grasp of, drawing techniques and perspective. Along with this, he showed a prodigious visual memory, which seemed able to take in the most complex buildings, or cityscapes, in a few seconds, and to hold them in mind, in the minutest detail—indefinitely, it seemed, and without the least apparent effort. Nor did the details need to be coherent, to be integrated into a conventional structure; among the most startling early drawings, Chris felt, were ones of demolition sites and earthquake scenes, with girders lying everywhere, exploded in all directions, everything in complete, almost random disarray. Yet Stephen remembered these scenes and drew them with the same fidelity and ease with which he drew classical models. It seemed to make no difference whether he drew from life or from the images in his memory. He needed no aide-mémoire, no sketches or notes—a single sidelong glance, lasting only a few seconds, was enough.

Stephen also showed abilities in spheres besides the visual. He was very good at mime, even before he was able to speak. He had an excellent memory for songs and would reproduce these with great accuracy. He could copy any movement to perfection. Thus Stephen, at eight, showed an ability to grasp, retain, and reproduce the most complex visual, auditory, motor, and verbal patterns, apparently irrespective of their context, significance, or meaning.

It is characteristic of the savant memory (in whatever sphere—visual, musical, lexical) that it is prodigiously retentive of particulars. The large and small, the trivial and momentous, may be indifferently mixed, without any sense of salience, of foreground versus background. There is little disposition to generalize from these particulars or to integrate them with each other, causally or historically, or with the self. In such a memory there tends to be an immovable connection of scene and time, of content and context (a so-called concrete-situational or episodic memory)—hence the astounding powers of literal recall so common in autistic savants, along with difficulty extracting the salient features from these particular memories, in order to build a general sense and memory. Thus the savant twins, calendrical calculators whom I described in *The Man Who Mistook His Wife for a Hat*, while able to itemize every event of their lives from about their fourth year on, had no sense of their lives, of historical change, as a whole. Such a memory structure is profoundly different from the normal and has both extraordinary strengths and extraordinary weaknesses. Jane Taylor McDonnell, author of *News from the Border: A Mother's Memoir of Her Autistic Son*, says of her son: "Paul doesn't generalize the particulars of his experience into the habitual, the ongoing, as many (most) other people do. Each moment seems to stand out distinctly, and almost unconnected with others, in his mind. So nothing seems to get lost, repressed, in the process." So it was, I often thought, with Stephen, whose life experience seemed to consist of vivid, isolated moments, unconnected with each other or with him, and so devoid of any deeper continuity or development.

Though Stephen would draw incessantly, he did not seem to take any interest in the finished drawings, and Chris might find them in the wastebasket or just left on a desk. Stephen did not even seem to concentrate on his subject while he was drawing. "Once," Chris related to me, "Stephen was sitting opposite the Albert Memorial: he was doing a fabulous picture of that, but at the same time looking all around—at buses, the Albert Hall, whatever."

Though he did not think that Stephen needed to be "taught," Chris devoted himself as much as possible to Stephen and his drawing, providing him with models, with encouragement. This was not always easy, because Stephen did not show much personal feeling. "In a way, he was responsive to us, the adults—he would say, 'Hullo, Chris,' or 'Hullo, Jean.' But it was difficult to reach him, to know what was in his mind." He seemed not to understand different emotions and would laugh if one of the children had a temper tantrum or screamed. (Stephen himself rarely had tantrums at school, but when he was little, he would sometimes have them at home.)

Chris was central in Stephen's life between 1982 and 1986. He would often take Stephen, along with his class, on outings in London, to see St. Paul's, to feed the pigeons in Trafalgar Square, to see Tower Bridge being raised and lowered. These outings finally incited Stephen to words in his ninth year. He would recognize all the buildings and places they passed, traveling in the school bus, and excitedly call out their names. (When he was six he had learned to ask for "paper" when he needed it—for many years, he had not understood how to ask for anything, even by gesture or pointing. This therefore was not only one of his first words, but the first time he understood how to use words to address others—the social use of language, something normally achieved by the second year of life.)

There were some fears that if Stephen began to acquire language he might lose his astounding visual gifts, as had happened, coincidentally or otherwise, with Nadia. But both Chris and Lorraine Cole felt that they had to do their utmost to enrich Stephen's life, to bring him from his wordless isolation into a world of interaction and language. They concentrated on making language more interesting, more relevant, to Stephen, by linking it with the buildings and places he loved, and got him to draw a whole series of buildings based on letters of the alphabet ("A" for Albert Hall, "B" for Buckingham Palace, "C" for County Hall, and so on, right up to "Z" for London Zoo).

Chris wondered if others would find Stephen's drawings as extraordinary as he did. Early in 1986, he entered two of them in the National Children's Art Exhibition; both were exhibited, and one of them won a prize. Around this time, Chris also sought an expert opinion on Stephen's abilities from Beate Hermelin and Neil O'Connor, psychologists who were well known for their work on autistic savants. They found Stephen one of the most gifted savants they had ever tested, immensely proficient in both visual recognition and drawing from memory. On the other hand, he did rather poorly in general intelligence tests, scoring a verbal IQ of only 52.

Word of Stephen's extraordinary talents started to spread, and arrangements were made to film him as part of a BBC program on savants, titled "The Foolish Wise Ones." Stephen took the filming very calmly, not at all fazed by cameras and crews—possibly even enjoying it slightly. He was asked to draw St. Pancras Station ("a very 'Stephen' building," as Lorraine Cole emphasized, "elaborate, detailed and incredibly complicated"). The accuracy of his drawing is attested by a photograph taken at the same time. (There is, however, a curious error: Stephen makes a mirror reversal of the clock and the whole top of the building.) His accuracy was astounding, as were the speed with which he drew, the economy of line,

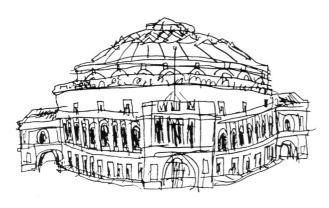

*"**A** is for Albert Hall"*

*"**U** is for Underground Train"*

Part of Stephen's *London Alphabet*, drawn when he was ten.

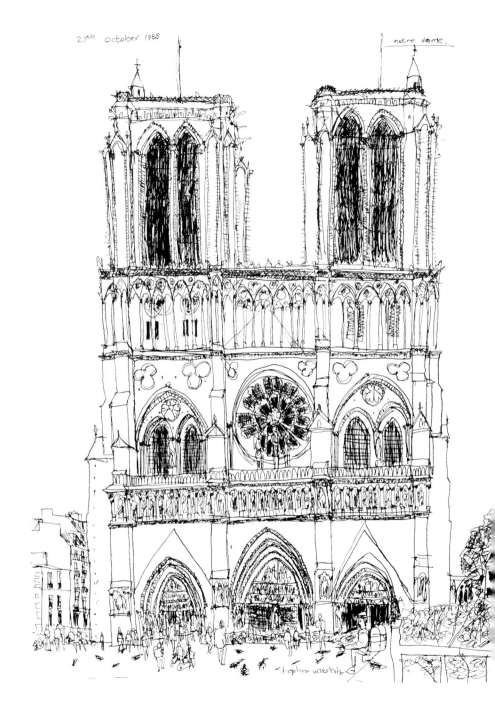

2th october 1988

notre dame.

Stephen wiltshire

Notre Dame, drawn when Stephen was fourteen.

Stephen's rendition of Matisse's *Dance* conflates the drawing of the Hermitage version with the colors of the Museum of Modern Art version.

A Matisse face (upper left), reproduced by Stephen directly, and then by memory at hourly intervals.

The old houses on the Herengracht in Amsterdam, as seen from Stephen's hotel window; and the Doge's Palace in Venice.

One of several drawings Stephen made of St. Basil's, in Red Square.

An aerial view of the Chrysler Building in New York, from the top
of the Pan Am Building.

A lavish interior at the Chicago Theater.

Three tiny sketches, done at speed: an Arizona landscape, an elephant at the London Zoo, and St. Basil's.

the charm and style of his drawings—it was these that won viewers' hearts. The BBC program was shown in February of 1987 and aroused a storm of interest—letters poured in, asking where Stephen's drawings could be seen, and publishers offered contracts. Very soon a collection of his work, to be called simply *Drawings*, was slated for publication; and it was this I received the proofs of, in June of 1987.

Stephen, only thirteen, was now famous throughout England—but as autistic, as disabled, as ever. He could draw, with the greatest ease, any street he had seen; but he could not, unaided, cross one by himself. He could see all London in his mind's eye, but its human aspects were unintelligible to him. He could not maintain a real conversation with anyone, though, increasingly, he now showed a sort of pseudosocial conduct, talking to strangers in an indiscriminate and bizarre way.

Chris had been away for some months in Australia and returned to find his young pupil famous—but, he thought, completely unchanged. "He recognized that he'd been on TV, and that he'd had a book published, but he didn't go overboard, as many children would have done. He wasn't affected; he was still the Stephen I knew." Stephen had not seemed to miss Chris too much during his absence, but seemed glad to see him back, said "Hullo, Chris!" with a big smile on his face.

None of this quite added up for me. Here was Stephen being exhibited as a significant artist—the former president of the Royal Academy of Arts, Sir Hugh Casson, had called him "possibly the best child artist in Britain"—but Chris and others, even the most sympathetic, seemed to see him as greatly lacking in both intellect and identity. The tests that had been given to him seemed to confirm the severity of his emotional and intellectual defect. Was there, nonetheless, a mental and personal dimension, a depth and sensibility, in him that could emerge (if nowhere else) in his art? Was not art, quintessentially, an expression of a personal vision, a self? Could one be an artist without having a "self"? All these questions had been in my mind since I had first seen Stephen's pictures, and I was eager to meet him.

The opportunity came in February of 1988, when Stephen came to New York, accompanied by Chris, to make another television documentary. Stephen had been in New York for a couple of days, seeing and drawing the sights of the city, and—his greatest thrill—flying over it in a helicopter. I thought he might like to see City Island, the little island off New York where I live, and invited him to come to my house. He and Chris arrived in the middle of a snowstorm. Stephen was a demure, grave little black boy, though clearly with an impish side. He looked young to me, closer to ten than thirteen, with a smallish head, tilted to one side. He reminded me somewhat of the autistic children I had seen before, with a head-nodding mannerism or tic, and some odd flapping movements of the hands. He never looked at me directly but seemed to glance at me, briefly, out of the corners of his eyes.

I asked him how he was finding New York, and he said "Very nice" with a strong Cockney accent. I have little recollection of his saying much else; he tended to be very quiet, almost mute. But his language had developed a good deal since the early days, and there were times, Chris said, when he would get excited and almost babble. He had been very excited on the plane—he had never flown before—and, Chris told me, "talked with the cabin crew and other passengers, showing his book around on the flight."[5]

Stephen wanted to show me his latest drawings, of New York—they were all in a portfolio Chris was carrying—and I admired them (especially the aerial ones he had done from the helicopter) as he showed them to me. He nodded emphatically as he displayed them, calling some of them "good" and "nice." He seemed to have no sense of either vanity or modesty, but showed me his drawings, commented on them, in an

---

[5] When Stephen was invited to sit in the jump seat for the New York landing, Chris recalled a prescient dream that he had reported before they left London. "I am being the pilot of the jumbo jet," Stephen had said. "I can see the skyscrapers and the Manhattan skyline."

ingenuous way and with a total absence of self-consciousness.

After he had shown me these, I asked him if he would draw something for me, perhaps my house. He nodded, and we wandered outside. It was snowing, cold and wet, not a day to linger. Stephen bestowed a quick, indifferent look at my house—there hardly seemed to be any act of attention—glanced then at the rest of the road and the sea at the end of the road, then asked to come in. As he took up his pen and started drawing, I held my breath. "Don't worry," Chris broke in, "you can talk at the top of your voice if you want to. It won't make any difference—you can't interrupt him—he could concentrate if the house was falling down." Stephen did not make any sketch or outline, but just started at one edge of the paper (I had a feeling he might have started anywhere at all) and steadily moved across it, as if transcribing some tenacious inner image or visualization. As he was putting in the porch railings, Chris remarked, "I didn't see any of that detail there."

"No," said Stephen, his expression implying, "No, you wouldn't."

Stephen had not studied the house, had made no sketches, had not drawn it from life, but had, in a brief glance, taken everything in, extracted its essence, seen every detail, held it all in his memory, and then, in a single, swift line, drawn it. And I did not doubt that, had we let him, he could have drawn the entire street.

Stephen's drawing was accurate in some ways, but took various liberties in others—he gave my house a chimney where there was none, but omitted the three fir trees in front of the house, the picket fence around it, and the neighboring houses. He focused on the house to the exclusion of anything else. It has often been said that savants have photographic or eidetic memories, but as I photocopied Stephen's drawing I thought how *unlike* a Xerox machine he was. His pictures in no sense resembled copies or photographs, something mechanical and impersonal—there were always additions, subtractions, revisions, and, of course, Stephen's unmistakable style. They were images and showed us some of the immensely complex neural processes that are needed to make a visual and graphic image. Stephen's drawings were individual constructions, but could they be seen, in a deeper sense, as creations?

His drawings (like those of my patient José) had a closeness to the actual, a literalness and naïveté. Clara Park, the mother of an autistic artist, has called this an "unusual capacity to render the object as perceived" (not *conceived*). She also writes of an "unusual capacity for delayed rendition" as characteristic of savant artists; this indeed was very striking in Stephen, who, after a single glance at a building, would retain it effortlessly for days or weeks, and then draw it as if from life.

Sir Hugh Casson wrote in his introduction to *Drawings*:

> Unlike most children, who tend to draw less from direct observation than from symbols or images seen second-hand, Stephen Wiltshire draws exactly what he sees—no more, no less.

Artists are full of symbols and images seen second hand and bring to their drawings not only the conventions of representation they acquired as children, but the entire history of Western art. It may be necessary to leave these behind, to leave behind even the primal category of "objecthood." As Monet put it:

> Whenever you go out to paint, try to forget what objects you have in front of you—a tree, a house, a field, or whatever. . . . Merely think, here is a little squeeze of blue, here an oblong of pink, here a streak of yellow, and paint it just as it looks to you, the exact color and shape, until it gives your own naive impression of the scene before you.

But Stephen (if Casson is right) and José, and Nadia and other savants, may not have to make such "deconstructions," may not have to relinquish such constructs, because (at many levels, from the neural to the cultural) they never made them in the first place or made them to a far smaller extent. In this way their situation is radically different from that of the "normal"—though this does not mean that they cannot be artists, too.

I started to wonder, too, about the relationships in Stephen's life: how important they were, to what extent they had developed, in the face of his autism (and devastating early loss). His relationship with Chris Marris, perhaps the most crucial during his last five years at Queensmill, had threatened to end when, in July of 1987, Stephen had to leave Queensmill for a secondary school. For a while, Chris had arranged to continue seeing Stephen on weekends, to take him on drawing outings around London, and even on his first trips to New York and Paris. But by May 1989 these expeditions had come to an end, and Stephen seemed to lack the initiative to do much drawing on his own. It seemed as if he needed another person to get him going, to "facilitate" his drawing. Whether he missed, or mourned, Chris in a more personal way was far less clear. When I later spoke of Chris to Stephen, he would talk about him (always as "Chris Marris" or "Mr. Marris") in a very flat and factual way, without any apparent emotion. A normal child would be deeply distressed at the loss of someone who had been so close for many years, but no such distress was apparent in Stephen. I wondered if he was repressing painful feelings, or distancing himself from them, but I was not sure whether, in his autistic way, he even had any personal emotion here at all. Christopher Gillberg writes of a fifteen-year-old autistic boy whose mother had died of cancer. Asked how he was doing, the boy replied, "Oh, I am all right. You see I have Asperger syndrome which makes me less vulnerable to the loss of loved ones than are most people." Stephen, of course, would never have been able to articulate his inner state in this way, and yet one had to wonder whether he took the loss of Chris with some of the same flatness as Gillberg's young patient—and whether such a flatness might not characterize most of the human relationships in his life.

Into this void irrupted Margaret Hewson. Margaret had been Stephen's literary agent since the BBC program two years before and had developed an increasing personal and artistic interest in him. I had first met her in 1988, when, with Stephen, we roved around London on a drawing expedition. Margaret and Stephen, it was evident to me, got on very well. Stephen, though perhaps incapable at this point of any depth of feeling or caring, nevertheless showed strong instinctive responses to different people. He had taken to Margaret from the start—attracted, I think, by her enormous energy and impetus, the exhilarating, whirlwind atmosphere she seemed to create all around her, and by her obvious feeling for him and his art. Margaret seemed to know everyone and have been everywhere, and perhaps this gave Stephen a sense of a larger world, of horizons far beyond the narrow ones that had confined his life hitherto. Margaret, finally, was very knowledgeable about art, a knowledge that extended from art history to the technical details of drawing.

In the fall of 1989, Margaret began obtaining drawing commissions for Stephen and taking him out drawing every weekend, along with her husband and partner in the literary agency, Andrew. She instantly abolished the use of tracing paper and rulers (such as he had used for some of the drawings in his second book, *Cities*, published in 1989), and insisted he draw freehand in ink. "One can learn the value of a line only by going straight into ink and making mistakes," she declared. Under Margaret's impetus and guidance, Stephen started to draw regularly once again, and to draw more boldly than he had ever done. (And yet even in *Cities* there had been some extraordinary freehand improvisations—imaginary cities, which Stephen had conceived, conflating the features of several real ones.)

After a morning of traveling and drawing, they would all return to the Hewson house for lunch, where they would often be joined by the Hewsons' daughter, Annie, only a few years older than Stephen. He seemed to look forward to these outings and would become excited on Sunday mornings, waiting for Margaret and Andrew to collect him. For their part, the Hewsons felt a real affection for Stephen, even though they were not sure he felt any actual affection for them. They started taking him on occasional longer excursions—a trip to Salisbury, and two weekends in Scotland.

Stephen's obvious fondness for the visual effects of water—he lived near a canal in London and would sometimes walk along it with his mother or sister and do little sketches of the boats and locks—suggested to Margaret a theme for a new book. Together they would visit cities built around canals, "floating cities"—Venice, Amsterdam, and Leningrad—and draw these.

Late in the fall of 1989, Margaret impulsively phoned up Mrs. Wiltshire and suggested that Stephen and his sister, Annette, come to Venice with them for their Christmas holiday. The trip went exceedingly well. Stephen, now fifteen, seemed to cope easily with the uncertainties of travel, which would have thrown him only a few years before. He portrayed, as Margaret hoped he might, St. Mark's, the Doge's Palace, the great monuments of Venetian culture, and obviously enjoyed drawing them. But when asked what he thought of Venice, after a week in this high point of European civilization, he could only say, "I prefer Chicago" (and this not because of its buildings but its American cars—Stephen had a passion for these and could identify, name, and draw every postwar model ever made in the United States).

A few weeks later, plans were made for his next trip, to Amsterdam. Stephen approved of the trip for a very specific reason: he had seen photographs of the city, and said, "I prefer Amsterdam to Venice because it has *cars.*" Once again, Stephen captured perfectly the feeling of the city, from his formal drawings of the Westerkerk and the Begijnhof to his tiny, charming sketch of an odd statue with a street organ. Stephen seemed very much alive, and in high humor, in Amsterdam and started to show new aspects of himself. Lorraine Cole, who came along on the trip, was particularly startled at the changes she saw:

> When he was little, nothing was amusing to Stephen. He now finds all manner of things funny and his laughter is incredibly infectious. He has gone back to caricaturing people around him, and he takes great pleasure watching his victims' reactions.

One evening in Amsterdam, when Stephen was due to give an interview for a television show, Margaret developed a severe attack of asthma and had to stay in her hotel room. Stephen was very distressed, refused to do the TV show, and could not be budged from the end of Margaret's bed. "I'm going to stay with you till you get better," he declared. "You're not going to die." Margaret and Andrew were very touched by this.

"This was the first time we saw that he cared," she told me.[6]

Was it possible that Stephen was starting to show some belated personal development, in spite of his autism? Intrigued by Margaret's report on the Amsterdam trip, I arranged to come along on the next visit, to Moscow and Leningrad, planned for May of 1990. I flew to London, met Stephen and Margaret there, and did some testing with Stephen. These tests, devised by Uta Frith and her colleagues, require one to react to various cartoons, some of which relate simple sequences of events while others cannot be understood without attributing different intentions, perspectives, beliefs, or states of mind (and sometimes dissemblings) to the characters involved. Stephen, it was clear, had a very limited ability to

---

[6] Visiting the autistic artist Jessy Park, I was struck by the great affection her parents showed for her. "I see how you love her," I said to her father. "Does she love you, too?"

"She loves us as much as she can," he replied.

imagine others' states of mind. (Frith writes that one researcher "carried out an informal survey in America using cartoons from *The New Yorker*. Very able and highly educated autistic people failed to understand them, or find them funny.")

I also gave him a large jigsaw puzzle, which he put together very swiftly. I then gave him a second puzzle, this time *face down*, so that he did not have the picture to assist him. He did this just as quickly as the first. The picture—meaning—it seemed, was not necessary to him; what was preeminent, and spectacular, was his ability to apprehend a large number of abstract shapes, and to see in a trice how they fitted together.

Such performances are characteristic of autistic people, who also excel in tests of block design and especially in finding embedded figures. Thus the psychologist Lynn Waterhouse, testing one visual savant, J.D. (who as a boy, his parents said, was able to complete a five-hundred-piece jigsaw puzzle in about two minutes and thereafter had to be given five-thousand-piece puzzles), found he performed "phenomenally well" on almost every visual-perception test she could give him: on tests of line orientation, visual gestalt closure, block design, and so on, he obtained nearly perfect results, in each case performing the tests at many times the normal rate. Stephen, like J.D., had prodigious powers of abstract-pattern recognition and visual analysis. But this alone could not explain his drawing—J.D., despite his perceptual powers, was not especially gifted in drawing.

Stephen, then, was calling on another sort of power, of vivid representation—representation that created an external form for his perceptions, and that bore a very recognizable and personal style. Whether this power of representation entailed any depth of inner resonance or response remained completely unclear.

Given Stephen's powers of abstract visual analysis, how important was "meaning" to him? How much did he get the meaning of what he drew? And how much did it matter whether he did or not? I showed Stephen a portrait by Matisse

Prodigies

Olivers sakKs house

Olivers sack's House

stephen Wilts

Oliver Sack's House

Stephen's first drawing of my house was done from memory, after taking a quick glance on his first visit in February 1988. The second, also from memory, was done more than two years later; the third a year after this. Although he has changed various details with the passage of time, he manages to extract the "style" of the house in each version.

and asked if he would draw it. (Margaret and Andrew are very fond of Matisse, and it was a print of theirs that I showed Stephen.) He drew it, from the original, swiftly and confidently; it was not wholly, literally accurate, but it was very Matisse-like. When I asked him to draw it again, from memory, an hour later, he drew it differently, and, another hour after this, yet differently again; but all his drawings (he did five in all), while different in detail, were strikingly evocative of the original. In some sense, therefore, Stephen had extracted the drawing's "Matisseness," had permuted it various ways, and had made this central in all his copies. Was this purely formal, cognitive, a matter of getting Matisse's "style" in a formulaic way—or was he responding, at a deeper level, to Matisse's vision, his sensibility and art?

I asked Stephen if he remembered my house, which he had visited more than two years before, and if he would draw it again for me. He nodded and again drew the house, but with various revisions. He now gave it one lower window instead of two; he removed a pillar from the porch and made the steps more prominent. He kept the (fictitious) chimney, and now he added a fictitious American flag on a tall flagstaff as well—I think he felt these as the ingredients of a formulary "American" house. Thus the Matisse and my house were conceived, and represented, in a variety of versions. In both cases, he got the style at once, and his later drawings were improvisations within this style.

After all this testing, I was still bewildered. Stephen seemed so defective, and so gifted, simultaneously; were his defects and his gifts totally separate, or were they, at a deeper level, integrally related? Were there qualities, like autistic literalness and concreteness, that might in some contexts be gifts, in others deficits? The tests also gave me a feeling of disquiet, as if I had spent days reducing Stephen to defects and gifts and not seeing him as a human being, as a whole. I had just reread Uta Frith's book *Autism: Explaining the Enigma* and wrote to her, "Tomorrow I go with Stephen to Russia. . . . I have seen something of his odd skills and defects—I have yet to see him as a mind and person. Perhaps a week of being with him will show me this."

With these hopes, then, I set out with Stephen for Russia. Sitting at Gatwick airport, waiting for our flight, I was impressed by his deep concentration. He sat enthralled with the magazine *Classic Cars.* He looked at the pictures with extraordinary intentness—he did not raise his head from the magazine for more than twenty minutes. Occasionally he bent closer to inspect a detail—what he saw, I thought, would be forever imprinted on his cortex. Once in a while, he suddenly laughed. What, in this abstract exercise, excited his amusement?

In flight, Stephen immersed himself in a drawing of Balmoral, after studying a postcard of the castle. He was oblivious of the conversations going on around him, the magnificent landscapes and seascapes below.

At Moscow airport, Stephen, very quiet, looked at the cars—yellow cabs and black Zils with license plates starting with "MK." A hideous smell of unrefined gas hung over the airport. Stephen sniffed, wrinkled his nose; he is extremely sensitive to smells. As we drove into the city, at 2 a.m., we saw tall, silvery birches by the side of the road and an immense, low moon. Even Stephen, seemingly oblivious of his surroundings before, gazed at the vast moonlit landscape with delight, his nose pressed against the cold window of the bus.

The next morning, as we walked around Red Square, Stephen was actively curious, taking snapshots, peering at buildings, struck by their novelty. Other people turned around and stared at him in the street—black people, apparently, were unusual in Moscow. He found a spot from which he wanted to draw the Spassky Tower and had Margaret set his stool in precisely this place. Not there, or there, but *here*—passive in so many ways, he was entirely master now. In the middle of Red Square, he was a tiny figure, wearing a fur cap and navy-blue woolen gloves. Dozens of tourists swarmed around in the brilliant May sun; many of them peered at Stephen's drawing.

Stephen ignored them, or was unconscious of them, and drew on undisturbed. He hummed to himself as he drew, holding his pen, characteristically, awkwardly, childishly, between his third and fourth fingers. At one point he broke into giggles and laughter—but this, it turned out, was because a scene in *Rain Man* ("Don't you *dare* drive!" he said) kept entering his mind. Margaret sat to one side as he drew, encouraging— "Good! Clever boy!"—advising him on aesthetic points and architectural details. At her suggestion, for instance, Stephen examined the tower's crenellation. She is almost a collaborator in a way, and though his talent is so personal and indigenous, he clearly looks to her for affectionate and always affirmative comments.

Later, we visited the History Museum, an eclectic red brick building, designed by an English architect. Margaret instructed Stephen, "Have a jolly good look at that building. Study it. Take in the vocabulary of that building now—I want you to draw it from memory afterward." But what Stephen actually drew later was different from the History Museum and bore half a dozen onion domes, not present in the original.

I first wondered whether this was a defect of memory and asked him if he would draw St. Basil's from memory. He did this instantly, a very accurate and quite charming sketch, in all of two minutes. Later in the day, he started a drawing of the vast shopping arcade at GUM, which he finished at leisure over a Coke in the hotel. He had retained by memory even the shop signs, although they were, to him, unintelligible Cyrillic letters. There was no faulting his memory, clearly.

Margaret and I tried to think what had happened with the History Museum; Stephen was distracted when asked to memorize it (the police in Red Square made him nervous) and when asked how he felt about it would only say, "It's all right" (which meant he did not like it). He tried to make it more attractive, I think, by crowning it with onion domes, but these were so out of keeping with the base that the resulting building looked hardly possible.

The next morning, as we met for breakfast in the hotel dining room, Stephen greeted me with a booming "Hullo, Oliver!" shouted with great friendliness and warmth, or so I thought. But then I was not sure—was it merely a social automatism? The great neurologist Kurt Goldstein wrote of another autistic boy:

> He becomes fond of some people. . . . At the same time, however, his emotional responses and human attachments remain shallow and perfunctory. Meeting him at intervals of several months, one is welcomed and bid goodbye with the same impersonal kindness as if contact were only real as long as it lasted during concrete presence . . . it is a presence without emotional content.

At an Intourist shop, I bought a piece of amber. Stephen glanced at it indifferently—it held no visual appeal for him—until I rubbed it and showed him how it became electrically charged. It attracted tiny pieces of paper now, so that when I put the amber a few inches away they suddenly flew up to it. His eyes opened wide in astonishment; he took the amber from me and repeated the electrification by himself. But then his wonder seemed to fade. He did not ask what happened or why, and he seemed uninterested when I explained it. I was excited at seeing his initial astonishment—I had never seen him truly astonished before—but then it faded, died out. And this, to me, seemed rather ominous.

At dinner, chortling, Stephen drew a cartoon of us all at the table, with himself fanning me. (I am sensitive to heat and always carry a Japanese fan, which he had often seen me use.) He portrayed me as cowering under the impact of the fan, and himself as large, powerful, in command—this was a symbolic representation, the first one I had seen him make.

Traveling, living with Stephen—we had now been together for five days—I became very conscious of how brittle he was physiologically, of the profound fluctuations in his state. There were times when he was animated and interested in his surroundings and could do brilliant, funny impersonations and cartoons; and there were times when he would revert to the deepest autism and respond, if at all, like an automaton, echolalically. Such fluctuations, usually lasting a few hours, rarely days, are common in children with classical autism, though their cause is not understood. They had been much worse, I was told, when Stephen was younger.

The next day we boarded a train for the daylong journey to Leningrad. Margaret had put together a huge hamper of provisions, more than enough for ourselves and any fellow passengers in the compartment. As the train got under way, we started with an early breakfast—we had left the hotel at five to make this early train. As she unpacked the basket, Stephen, half convulsively, swooped his head and sniffed everything as it came out. I was reminded of some of my postencephalitic patients, and some people with Tourette's syndrome, whom I had also seen with olfactory behaviors of this sort. I suddenly realized that Stephen's smell-world might be as vivid as his visual one; but we do not have the language, the means, to convey such a world.

Stephen gazed uncertainly at our hard-boiled eggs—was it possible that he had never cracked one open? Playfully, I took one and cracked it on my head; Stephen was delighted and burst out laughing. He had never seen a hard-boiled egg cracked in this way, and he gave me a second egg to see if I would do it again, and then, reassured, cracked one on his own head. There was something spontaneous in this egg cracking, and I think Stephen felt easier with me after this, because I had shown him how playful, how silly, I could be.

After breakfast, Stephen and I played some word games. He was quite good at I Spy, and when I prompted him with "I spy with my little eye something beginning with 'c,' " he quickly reeled off "Coat, cat, café, coffee, cool, cup, cigarette." He was very good at filling in letters in incomplete words. And yet, at sixteen, he was still unable, despite repeated demonstrations, to judge the constancy of volume, despite differing heights, in different vessels—a concept that, as Piaget showed, most children grasp at seven.

The train passed through tiny villages of wooden houses and painted churches, giving me the sense of a Tolstoyan world, unchanged in a hundred years. As Stephen watched all this intently, I thought of the thousands of images he must be registering, constructing—all of which he could convey in vivid pictures and vignettes, but none of them, I suspected, synthesized into any general impression in his mind. I had the feeling that the whole visible world flowed through Stephen like a river, without making sense, without being appropriated, without becoming part of him in the least. That though he might retain everything he saw, in a sense, it was retained as something external, unintegrated, never built on, connected, revised, never influencing or influenced by anything else. I thought of his perception, his memory, as quasi-mechanical—like a vast store, or library, or archive—not even indexed or categorized, or held together by association, yet where anything might be accessed in an instant, as in the random-access memory of a computer. I found myself thinking of him as a sort of train himself, a perceptual missile, traveling through life, noting, recording, but never appropriating, a sort of transmitter of all that rushed past—but himself unchanged, unfed, by the experience.

As we approached Leningrad, Stephen decided it was time to draw. "Pencil, Margaret, dahhling!" he said. I was amused by the "dahhling," a Margaretism that he had adopted, and I could not decide whether it was automatic or more conscious, a humorous parody. The train was very jolty, and I was able to make only brief notes. But Stephen was perfectly able to draw, with his usual speed and fluency; I had been amazed by this earlier, on the airplane. (He looked clumsy, but he picked up some motor skills, it seemed, almost instantly, as some autistic people seem to do. In Amsterdam, he had had no hesitation in walking a narrow gangway to a houseboat, something he had never done before, and this reminded me of another autistic youngster I had met, who suddenly walked a tightrope, expertly and fearlessly, the day after seeing it done at the circus.)

Finally, after eleven hours of slow traveling—rural Russia slowly unrolling before us—we arrived at a grand station in Leningrad, a station of faded, prerevolutionary, czarist splendor. The whole panorama of the city, with its fine, low, eighteenth-century buildings, its sense of European cosmopolitan civilization, could be seen from our hotel windows, glittering in the northern white night. Stephen was eager to see it in full daylight and decided he would draw it the next morning, first thing. I was not in the room when he started, but Margaret told me later that he made an interesting false start. There was a famous old cruiser, the *Aurora*, moored in the Neva, and Stephen had drawn it way out of proportion to the buildings on the other side. When he realized what he had done, he said, "I'll just start again. It's no good. It won't work." He tore off another sheet of paper and started again.

The flagrant incongruity, initially, between boat and buildings made me think of other, smaller incongruities in his work, the fact that he might use multiple perspectives in his drawings and that these did not always precisely coincide.[7]

Later that day, we went to the Alexander Nevsky Monastery and found ourselves, unexpectedly, in the middle of a Russian Orthodox wedding. The choir consisted of a gaunt, ragged huddle, led by a blind woman with blazing blue eyes. But their voices were marvelous, almost beyond bearing, especially that of the basso profundo, who looked, Margaret and I felt, like an escapee from the Gulag. Margaret thought that Stephen was unaffected by their voices, but I felt the opposite, that he was profoundly affected—a measure of how difficult it was, at times, to know *what* he was feeling.

The climax of our time in Leningrad was a visit to the Her-

---

[7] This was pointed out to me, with many examples, by a very acute correspondent, John Williamson, of Brownsville, Texas, who plans to write about them at length.

mitage, but Stephen showed a somewhat childish reaction to the incredible paintings there. "See how it's built up in blocks?" Margaret said of one Picasso, a woman with a tilted head. Stephen merely asked, "Has she got a pain?"

Margaret told Stephen to take special note of the Matisse *Dance*, and Stephen gazed at it, without much sign of interest, for a full thirty seconds. Back in London, Margaret suggested he draw it, and he did—unhesitatingly, brilliantly. It was only later that a curious conflation was noted (again by the observant Mr. Williamson): Stephen had used the forms of the dancers in the Hermitage painting but had given them the colors of another version of the painting (which hangs in the Museum of Modern Art in New York). His sister, Annette, it turned out, had given him a poster of the MOMA *Dance* years before, and now he gave the "American" colors to the "Russian" picture. One might wonder whether this was a lapse of memory or a confusion, but Stephen, I suspect, was being playful, and *decided* to give the Hermitage picture the MOMA colors, as he decided to give the History Museum onion domes (or, for that matter, my house a chimney, or, in another drawing, the Rockefeller Center Prometheus a penis).

Weary from a day of touring and drawing, we left the Hermitage and headed back to our hotel for tea. Seeing that Stephen needed some diversion, Margaret said to him, "*You* be the teacher now. . . . You, Oliver, the pupil."

A glint appeared in Stephen's eye. "What is two take away one?" he asked.

"One," I said promptly.

"Good! Now twenty minus ten?"

I pretended to think for a bit, then said, "Ten."

"That's very good," Stephen said. "Now sixty minus ten?"

I cogitated hard, screwed up my face. "Forty?" I said.

"No," said Stephen. "Wrong. Think!"

I tried to help myself by holding up my fingers in multiples of ten. "I've got it—fifty."

"Right," said Stephen, with an approving smile. "Very good. Now, forty minus twenty."

That was really difficult. I thought for a full minute. "Ten?"

"No," said Stephen. "You must concentrate! But you did pretty well," he added kindly.

The episode was a stunning imitation of an arithmetic lesson such as one might give to a retarded child. Stephen's voice, his gestures, mimicked to perfection those of a well-meaning but condescending teacher, specifically (I felt with some discomfort) mine when I had tested him in London. He had not forgotten this. It was a lesson to me, to all of us, never to underestimate him. Stephen delighted in reversing roles, just as in his cartoon of himself fanning me.

The Russia trip was in some ways delightful, exciting, in others saddening, disappointing, disillusioning. I had hoped to get behind Stephen's autism, to see the person underneath, the mind; but there had been only the merest intimation of this. I had hoped, perhaps sentimentally, for some depth of feeling from him; my heart had leapt at the first "Hullo, Oliver!" but there had been no follow-up. I wanted to be liked by Stephen, or at least seen as a distinct person—but there was something, not unfriendly, but de-differentiating in his attitude, even in his indifferent, automatic good manners and good humor. I had wanted some interaction; instead, I got a slight sense, perhaps, of how parents of autistic children must feel when they find themselves faced with a virtually unresponsive child. I had still, in some sense, been expecting a relatively normal person, with certain gifts and certain problems—now I had the sense of a radically different, almost alien mode of mind and being, proceeding in its own way, not to be defined by any of my own norms.

Yet there were times—the egg cracking, the pupil-and-teacher game together—when I felt a current between us, so I still hoped for some sort of relationship with him and made a point of visiting him each time I went to London, generally a few times a year. On one or two occasions I was able simply to go for a walk with Stephen. I hoped, still, that he might unwind, show me something of his spontaneous, "real" self. But though he would always greet me with his cheery "Hullo, Oliver!" he remained as courteous, as grave, as remote as ever.

There was, however, one enthusiasm we shared—a fondness for car spotting. Stephen especially liked the grand convertibles of the 1950s and 1960s. My favorite cars, by contrast, were the sports cars of my youth—Bristols, Frazer-Nashes, old Jags, Aston Martins. Between us we could identify most of the cars on the road, and Stephen, I think, came to see me as an ally or comrade in the game of car spotting—but this was as close as we ever got.

Floating Cities was published in February of 1991, and quickly went to the top of the best-seller list in England. Stephen was told this, and said, "Very nice!" He seemed unaffected or uncomprehending, and that was the sum of it. He was, at this point, going to a new technical school, learning to be a cook, taking public transport, and beginning to acquire some of the skills of independent life. But Sundays remained consecrated to drawing, and his work, commissioned and uncommissioned, multiplied each weekend.

The question of Stephen's artistic talents often reminded me of Martin, a retarded musical and mnemonic savant whom I saw in the 1980s. Martin loved operas—his father had been a famous opera singer—and could retain them after a single hearing. ("I know more than two thousand operas," he once told me.) But his greatest passion was for Bach, and I thought it curious that this simple man should have such a passion. Bach seemed so intellectual, and Martin was a retardate. What I did not realize—until I started bringing in cassettes of the cantatas, of the Goldberg Variations, and once of the Magnificat—was that, whatever his general intellectual limitations, Martin had a musical intelligence fully up to appreciating all the structural rules and complexities of Bach, all the intricacies of contrapuntal and fugal writing; he had the musical intelligence of a professional musician.

I had never before properly recognized the cognitive structure of savant talents. I had, by and large, taken them to be an expression of rote memory and little else. Martin, indeed, had a prodigious memory, but it was clear that this memory, in relation to Bach, was structural or categorical (and specifically architectonic)—he *understood* how the music went together, how this variation was an inversion of that, how different voices could take up a line and combine them in a canon or fugue, and he could construct a simple fugue himself. He knew, for at least a few bars ahead, how a line would go. He could not formulate this, it was not explicit or conscious, but there was a remarkable *implicit* understanding of musical form.

Having seen this in Martin, I could now see analogues in the artistic, calendrical, and calculating savants I had also worked with. All of them had a genuine intelligence, but intelligence of a peculiar sort, confined to limited cognitive domains. Indeed, savants provide the strongest evidence that there can be many different forms of intelligence, all potentially independent of each other. The psychologist Howard Gardner expresses this in *Frames of Mind*:

> In the case of the *idiot savant* . . . we behold the unique sparing of one particular human ability against a background of mediocre or highly retarded human performances in other domains. . . . the existence of these populations allows us to observe the human intelligence in relative—even splendid—isolation.

Gardner postulates a multitude of separate and separable intelligences—visual, musical, lexical, etc.—all of them autonomous and independent, with their own powers of apprehending regularities and structures in each cognitive domain, their own "rules," and probably their own neural bases.[8]

---

[8] In a rare congenital condition, Williams syndrome, there is astonishing verbal (and social) precocity, combined with intellectual (and visual) defects—an extreme scatter between different intelligences. The combination of linguistic giftedness with intellectual deficiency is especially startling: children with Williams syndrome often appear exceptionally self-possessed, articulate, and witty, and only gradually is their mental deficit borne in on one. The precise neuroanatomical correlates of this are being investigated by Ursula Bellugi and others.

In the early 1980s this notion was put to the test by Beate Hermelin and her colleagues, exploring the powers of many different forms of savant talents. They found that visual savants were far more efficient than normal people at extracting the essential features from a scene or design, and at drawing these, and that their memory was not photographic or eidetic, but, rather, categorical and analytic, with a power to select and seize on "significant features," using these to build their own images.

It was also evident that once a structural "formula" had been extracted, it could be used to generate permutations and variations. Hermelin and her colleagues, along with Treffert, also worked with the blind, retarded, and enormously gifted musical prodigy Leslie Lemke, who, like Blind Tom a century ago, is as renowned for his improvisational powers as for his incredible musical memory. Lemke catches the style of any composer, from Bach to Bartók, after a single hearing, and can thereafter play any piece or improvise, effortlessly, in that style.

These studies seemed to confirm that there were indeed a number of separate, autonomous cognitive powers or intelligences, each with its own algorithms and rules, precisely as Gardner had hypothesized. There had been a certain tendency before this to see savant abilities as extraordinary, as freakish; but now they seemed to be brought back into the realm of the "normal," differing from ordinary abilities only by being isolated and heightened in degree.

But do savant powers really resemble normal ones? One cannot have contact with a Stephen, a Nadia, a Martin, with any savant, without sensing something deeply other in action. It is not just that savant performances are off the scale, statistically, or incredibly precocious in their first appearance (Martin could sing bits of operas before he was two)—but that they seem to deviate radically from normal developmental patterns. This was particularly clear with Nadia, who seemingly skipped the normal scribbling, schematic, and tadpole stages, and when she drew did so in a way unlike any normal child. So it was with Stephen, who at seven, we know from Chris, did "the most unchildlike drawings" he had ever seen.

The other side of the prodigiousness and precocity, the unchildlikeness, of savant gifts is that they do not seem to develop as normal talents do. They are fully fledged from the start. Stephen's art at seven was clearly prodigious, but at nineteen, though he may have developed a bit socially and personally, his talent itself had not developed too greatly. Savant talents in some ways resemble devices, ready-made, pre-set, and ready to go off. And this is how Gardner speaks of them: "Assume that the human mind consists of a series of highly tuned computational devices . . . and that we differ vastly from one another in the extent to which each of these devices is 'primed' to go off."

Savant talents, further, have a more autonomous, even automatic quality than normal ones. They do not seem to occupy the mind or attention fully—Stephen will look around, listen to his Walkman, sing, or even talk while he is drawing; Jedediah Buxton's huge calculations moved ahead at their own fixed, imperturbable rate while he went on with his daily life. Savant talents do not seem to connect, as normal talents do, to the rest of the person. All this is strongly suggestive of a neural mechanism different from that which underlies normal talents.

It may be that savants have a highly specialized, immensely developed system in the brain, a "neuromodule," and that this is "switched on" at particular times—when the right stimulus (musical, visual, whatever) meets the system at the right time—and immediately starts to operate full blast. Thus, for the twin calendar savants, seeing an almanac at the age of six set off their extraordinary calendrical skill—they were able, straightaway, to see large-scale structural regularities in the calendar, perhaps to extract unconscious rules and algorithms, to see how the correspondence of dates and days could be predicted, which the rest of us, if we could do at all, could do only with consciously worked out algorithms and a great deal of time and practice.

The converse of this sudden kindling or turning on is also seen on occasion in the sudden disappearance of savant talents, whether in retarded or autistic savants, or normal individuals with savant capacities. Vladimir Nabokov possessed, in addition to his many other talents, a prodigious calculating gift, but this disappeared suddenly and completely, he wrote, following a high fever, with delirium, at the age of seven. Nabokov felt that the calculating gift, which came and went so mysteriously, had little to do with "him" and seemed to obey laws of its own—it was different in kind from the rest of his powers.

Normal talents do not come and go in this way; they show development, persist, enlarge, take on a personal style as they establish connections, and embed themselves, increasingly, in the mind and personality. They lack the peculiar isolation, uninfluenceability, and automaticity of savant talents.[9]

But a mind is not just a collection of talents. One cannot maintain a purely composite or modular view of the mind, as many neurologists and psychologists now do. This removes that general quality of mind—call it reach or range or size or spaciousness—that is always instantly recognizable in normal people. It is a capacity that seems to be supramodal, and that shines through whatever particular talents there are. This is

---

[9] It is possible for savant and normal talents to coexist, sometimes in separate spheres (as with Nabokov); sometimes, confusingly, in the same sphere. I have had this impression strongly with an extremely gifted young man I have known since infancy. At two, Eric W. could read fluently—but this was not just hyperlexia; he read with comprehension. At the same age he could repeat any melody he heard, harmonize in singing with it, and had a grasp of fugue and counterpoint. By three he was doing remarkable drawings with perspective. At ten he wrote his first string quartet. He showed great scientific powers in early adolescence, and now, in his early twenties, is doing fundamental work in chemistry. (I never had any sense of Eric W. being autistic—he was full of spontaneity and playfulness as a child, and is full of deep feeling as an adult.) Had he had only savant talents, they would not have been capable of significant development or integration. Had he had only normal talents (at least in the graphic sphere) they would not have been presented in such a savantlike fashion. He has been singularly fortunate in having both.

what we mean when we say that someone has "a fine mind." A modular view of the mind, no less importantly, also removes the personal center, the self, the "I." Normally, there is a cohering and unifying power (Coleridge calls it an "esemplastic" power) that integrates all the separate faculties of mind, integrates them, too, with our experiences and emotions, so that they take on a uniquely personal cast. It is this global or integrating power that allows us to generalize and reflect, to develop subjectivity and a self-conscious self.

Kurt Goldstein was especially interested in such a global capacity, which he referred to as the organism's "abstract-categorical capacity," or "abstract attitude." Part of Goldstein's work was concerned with the effects of brain damage, and he found that whenever there was extensive damage, or damage involving the frontal lobes of the brain, there tended to be, over and above the impairments of specific abilities (linguistic, visual, whatever), an impairment of abstract-categorical capacity—often as damaging as, sometimes far more damaging than, the specific impairments. Goldstein also explored various developmental problems and (with his colleagues Martin Scheerer and Eva Rothmann) published the deepest study ever made of an idiot savant. Their subject, L., was a profoundly autistic boy, with remarkable musical, "mathematical," and memorial talents. In their 1945 paper "A Case of 'Idiot Savant': An Experimental Study of Personality Organization," they comment on the limitations of a multifactorial, or composite, theory of mind:

> [If] there exists . . . only a composite of individual capabilities which are so independent from each other . . . L. should have theoretically been able to become a proficient musician and mathematician. . . . Since this contradicts the facts of the case, we have to explain [why he did not] . . . despite his "interests" and "training."

He did not, they conclude, because, for all his impressive and real talents, there was something else, something global, irremediably missing:

L. suffers from an impairment of abstract attitude affecting his total behaviour throughout. This expresses itself in the linguistic sphere by his "inability" to understand or to use language in its symbolic or conceptual meaning; to grasp or formulate properties of objects in the abstract . . . to raise the question "why" regarding real happenings, to deal with fictitious situations, to comprehend their rationale. . . . The same impairment underlies his lack of social awareness and of curiosity in people, his limited values; his inability to register or absorb anything of the socio-cultural and interhuman matrix around him. . . . The same impairment to abstract is evidenced in his [savant] performance . . . [which] cannot be lifted out of its concrete context for reflection and verbalization. . . . Owing to his impaired abstract attitude, L. cannot develop his endowment, actively and creatively. . . . [It remains] *abnormally* concrete, specific and sterile; it cannot become integrated with a broader meaning of the subject, nor with social insight. . . . [It] approaches rather a caricature of a normal talent.

If Goldstein's formulations about idiot savants and autism are generally valid, and if Stephen is indeed lacking, or relatively lacking, in abstract attitude, how much of an identity, or a self, might he be able to acquire? What power of reflective consciousness might be possible for him? To what extent can he learn or be influenced by personal or cultural contact? To what extent can he make such contact? How much can he develop a genuine sensibility or style? How much is any personal (as opposed to technical) development possible for him? What might be the resonances of all this for his art? These and many other questions, which one encounters with the paradox of an immense talent attached to a relatively rudimentary mind and identity, become sharper in the light of Goldstein's considerations.

In October 1991, I met Stephen in San Francisco. I was struck by how much he had changed since I last saw him—

now seventeen, he was taller, handsomer, and his voice deeper. He was excited to be in San Francisco and kept describing the scenes he had seen on television of the 1989 earthquake, in short, haiku-like phrases: "Bridges snapped. Cars crushed. Gas bursting. Hydrants flowing. Gaps opening. People flying."

On the first day, we climbed to the top of Pacific Heights. Stephen started drawing Broderick Street, which snakes up to the top of the hill. He looked around vaguely while he was drawing, but was mostly engrossed in listening to his Walkman. We had asked him earlier why Broderick snaked, instead of going straight up. He could not say, or see, that it was because of its steepness, and when Margaret said "steep" to him, he just repeated it, echolalically. He still seemed clearly retarded or cognitively defective.

As we walked, we came upon a sudden enchanting revelation of the bay, dotted with ships, and with Alcatraz set like a gem in the middle. But, for a moment, I did not "see" this, I did not see a scene at all, just an intricate pattern of many colors, a highly abstract, uncategorized mass of sensations. Was this how Stephen saw it?

Stephen's favorite building in San Francisco was the Transamerica Pyramid. When I asked him why, he said, "Its shape," and then, with an uncertain air, "It's a triangle, an isosceles triangle . . . I like that!" I was struck by the fact that Stephen, with his often primitive language, should use the word "isosceles"—though it is typical of autistic people, sometimes in early childhood, that they may acquire geometrical concepts and terms to a far greater degree than personal or social ones.[10]

---

[10] Freeman Dyson, who has known Jessy Park since she was a child, remarks:

I've always felt she was the closest I would ever come to an alien intelligence. Autistic children are so strange and so different from us—and yet you can communicate; there are many things you can talk with her about. . . . [But] she has no concept of her own identity, she doesn't understand the difference between "you" and "I"—she uses pronouns almost indiscriminately. And so her universe is radically different from mine. Concrete social relations are for

He has very little explicit understanding of autism—this came out in an unlikely incident on Polk Street. We had, by a million-to-one chance, got behind a car with a license plate that spelled "autism." I pointed it out to Stephen. "What does that say?" I asked. He spelled it out, laboriously, "A-U-T-I-S-M-2."

"Yes," I said, "and that reads?"

"U . . . U . . . Utism," he stuttered.

"Almost, not quite. Not utism—autism. What is autism?"

"It's what's on that license plate," he answered, and I could get no further.

Clearly, he recognizes that he is different, that he is special. He has a veritable passion for *Rain Man* and, one must suspect, identifies with the Dustin Hoffman character, perhaps the only autistic hero ever widely portrayed. He has the entire soundtrack of the film on tape and plays it continually on his Walkman. Indeed, he can recite large portions of the dialogue, taking every part, with perfect intonation. (His preoccupation with the film and his constant playing of the cassette have not distracted him at all from his art—he can draw wonderfully even though his attention seems to be elsewhere—but it has made him far less accessible to conversation and social contact.)

Going along with Stephen's obsession with *Rain Man* is his fervent desire to visit Las Vegas. He wanted, when we got there, to spend time in a casino, as Rain Man had, and not, in his usual way, to see the buildings in town. So we spent a single night there and then, in a 1991 Lincoln Continental, set out across the desert, for Arizona. "He would have preferred a 1972 Chevrolet Impala," Margaret told me, but this, to Stephen's disappointment, was not available.

---

her very, very difficult to comprehend. On the other hand, with anything abstract, she has no trouble. So mathematics, of course, is no problem for her, and we can talk very easily about mathematics. . . . I think autism comes about as close as possible to the central problem of exploring the neurological basis of personality. Because these are people whose intelligence is intact, but something at the center is missing.

We pulled up to a parking lot near the Grand Canyon—part of the canyon was visible from here, but Stephen's attention was immediately distracted by the other cars in the lot. When I asked what he thought of the canyon, he said, "It's very, *very* nice, a very nice scene."

"What does it remind you of?"

"Like buildings, architecture," Stephen answered.

We found a spot for Stephen to draw the North Rim of the canyon. He started to draw, less fluently and assuredly, perhaps, than he would draw a building; but he seemed to extract the basic architecture of the rocks nonetheless. "You're a genius, Stephen," Margaret remarked.

Stephen nodded, smiled. "Ya, ya."

Knowing Stephen's love of aerial views, we decided to fly over the Grand Canyon in a helicopter. Stephen was excited and kept craning his head in all directions as we flew low through the canyon, skimming the North Rim, and then higher and higher to get a bird's-eye view of the whole. Our pilot kept talking about the geology and history of the canyon, but Stephen ignored him, and, I think, saw only shapes—lines, boundaries, shadows, shadings, colors, perspectives. And I, sitting next to him, following his gaze, started, I imagined, to see it through his eyes, relinquishing my own intellectual knowledge of the rock strata below, and seeing them in purely visual terms. Stephen had no scientific knowledge or interest, could not, I suspect, have grasped any of the concepts of geology, and yet such was the force of his perceptual power, his visual sympathy, that he would be able to get, and later draw, the canyon's geological features with absolute precision, and with a selectivity not to be obtained in any photograph. He would get the canyon's feel, its essence, as he had got the essence of the Matisse.

We set out across the desert once again, and as we climbed toward Flagstaff, the saguaros grew rarer—the last one, a bold loner, stood out at twenty-eight hundred feet. The bleak Bradshaw Range, where silver and gold were found in the eighties, rose to our left. We entered a flat plain covered with prickly pear, with occasional cattle roaming. Horses and burros, and occasionally pronghorn antelope, still roam these plains. The San Francisco Peaks floated high, like vast ships, on the horizon.

"Very nice landscape to put motorcars into," Stephen remarked. (He had earlier drawn a big green Buick against a backdrop of Monument Valley.) I was amused—and outraged: faced with the sublimest, grandest vista on the planet, Stephen could only think to put motorcars into it!

While I scribbled, Stephen drew cacti; he had seized on them as an emblem of the West, as he had seized on gondolas for Venice, skyscrapers for New York. An animal, probably a rabbit, darted across the road in front of us. Something got into me, and impulsively I cried, "Coypu!" Stephen was taken by the word, its acoustic contours, and repeated it with obvious pleasure a number of times.

The Arizona trip showed us that Stephen could get desert, canyons, cacti, natural scenes, in the same uncanny way as he could get buildings and cities. Most startling of all, perhaps, was an afternoon at the Canyon de Chelly, which Stephen descended with a Navajo artist, who showed him a special, sacred vantage point from which to draw and plied him with the myths and history of his people, how they had lived in the canyon centuries before. Stephen was indifferent to all this but went ahead in his nonchalant way—looking around, muttering and humming to himself—while the Navajo artist sat, hardly moving, consecrated to the act of drawing. And yet, despite their so different attitudes, Stephen's drawing was manifestly the better and seemed (even to the Navajo artist) to communicate the strange mystery and sacredness of the place. Stephen himself seems almost devoid of any spiritual feeling; nonetheless he had caught, with his infallible eye and hand, the physical expression of what we, the rest of us, call the "sacred."

Did Stephen somehow imbibe a sense of the sacred and project this into his drawing, or do we, looking at his drawing, project this ourselves? There was often disagreement between Margaret and myself as to what Stephen actually felt, as with the wedding music at the monastery in Leningrad. But here, in the Canyon de Chelly, our roles were reversed: Margaret felt that Stephen had indeed been awed by the sacredness of the place, while I was skeptical. This deep uncertainty about what Stephen actually thinks and feels comes up constantly, with everyone who knows him.

I sometimes wondered whether "emotion" or "emotional response" might be radically different in Stephen: no less intense, but somehow more localized than in the rest of us—object-bound, scene-bound, event-bound, without ever coalescing or extending into anything more general, without becoming a part of him. I sometimes felt that he picked up the mood or the atmosphere of places, people, scenes, by a sort of instant sympathy or mimicry, rather than through what would usually be called a sensibility. Thus he might echo, or reproduce, or reflect, the world's beauties, yet not have any "aesthetic sense." He might resonate to the "holy" atmosphere in the Canyon de Chelly, or in the monastery, and yet not have any "religious" sense of his own.

Back in our hotel, in Phoenix, I heard sounds of wind instruments coming from Stephen's room, next door. I knocked at his door and entered—Stephen was alone, his hands cupped around his mouth. "What was that?" I asked.

"A clarinet," he said, and then did a tuba, a saxophone, a trumpet, and a nose-flute, all with uncanny accuracy.

I returned to my room, thinking about Stephen's disposition and power to reproduce, its many levels, and how it dominated his life. As a child he had shown echolalia when spoken to, echoing the last word or two of whatever other people said, and this still occurred, typically when he was tired or regressed. Echolalia carries no emotion, no intentionality, no "tone" whatever—it is purely automatic and may even occur during sleep. Stephen's "coypu" the day before was more complex than this, for he had savored the sound, the peculiar emphasis I gave it, but did it in his own way, an imitation, with variations. Then, at a still higher level, there was his reproduction of *Rain Man*, in which he reproduced or represented entire characters, their interactions, conversations, and voices. He often seemed nourished and stimulated by these, but at other times taken over, possessed and dispossessed, by them.

Such a "possession" may occur at many levels and may also be seen in people with postencephalitic syndromes or Tourette's syndrome. An automatic mimicry can occur in these, a reflection of a low-level physiological force overriding a normal mind and personality. Such a force may determine the more automatic aspects of autistic mimicry, too. But there may also be, at higher levels, a sort of identity hunger—a need to take off, take on, take in, other personas. Mira Rothenberg has sometimes compared autistic people, in this sense, to sieves, constantly sucking in other identities but unable to retain and assimilate them. Yet, she points out, after thirty-five years of experience, she still feels there is always a real self that she can connect to in the autistic.

Our last morning in Phoenix, I was up at seven-thirty, watching the sunrise from my hotel-room balcony. I heard a cheery "Hullo, Oliver!" and there was Stephen on an adjacent balcony.

"Wonderful day," he said, and then, holding his yellow camera, snapped me as I smiled back from my balcony. This seemed such a friendly, personal act—it would stay in my mind as our farewell to Arizona. As we walked outside, he went over to the cacti: "Bye, Saguaro! Bye, Barrel! Bye, Prickly Pear, see you next time!"

The paradox of Stephen's art was sharpened for me, but without resolution, by this trip. Margaret was constantly delighted by his work and would hug him and say, "Stephen! You give such delight! You have no idea how much pleasure you give!" Stephen would give his goofy smile and chortle— but Margaret was right. He did, through his drawings, bring others great pleasure, and yet it was not clear that they were

associated in him with any emotion whatever, other than the
pleasure of a faculty being exercised and used.

At one point on our Arizona trip, stopping at a Dairy
Queen, Stephen ogled two girls sitting at a table and was so
fascinated by them, indeed, that he forgot to go to the rest
room. In some ways, he is a normal adolescent boy; neither
his autism nor his savantism precludes this. Later, he went up
to the girls—he is not unpersonable on first impression. But
he spoke to them in a manner so inappropriate and childlike
that they looked at each other, giggled, and then ignored him.

Adolescence, both physical and psychological, perhaps
slightly belated, now seems to be rushing ahead with great
speed. Suddenly, Stephen has developed a strong interest in
his appearance, his clothes, rock music, and girls. He never
seemed to notice mirrors when he was younger, Margaret said,
but now he is always checking himself, preening before them.
He has developed very decided tastes in clothing: "I like
western-style jeans, light blue, garment washed, and shirts . . .
and black western boots."

"What do you think of Oliver's shoes?" Margaret asked
archly on one occasion.

"Boring," he said, throwing a glance at them.

Very little social life, as yet, is possible for Stephen. He
meets people, superficially, but does not know how to talk
with them and has few friends or real relationships outside his
own family or the Hewsons. He is very close to his sister,
Annette, and can be affectionate to her. He feels himself the
man of the house, a protector of his mother; and he feels that
Margaret is very much a protector of himself. But for the most
part he is thrown back on his drawings, and on increasingly
charged and detailed daydreams.

The world that really excites Stephen at this point is that of
"Beverly Hills, 90210," a television show he adores. Last year,
I asked him about it: "I love Jennie Garth," he said. "She's the
coolest girl in L.A. She's got red lipstick. . . . She's twenty-one
years old. She's from Illinois. She's in 'Beverly Hills, 90210.' I
fell in love with Jennie Garth. It started in 1991, I think. She

plays Kelley Taylor. She always wears jeans and western-style shirts and bodysuits." It is not just Jennie Garth but the entire cast of the show that Stephen is in love with, and whom he now incorporates in more and more elaborate fantasies. "I collect their pictures," he said. "I sent them several drawings." Now he wants to design a penthouse for them on Park Avenue. They will all live together, and he will live with them, as "artist-in-residence." He will decide who may visit them and who may not. In the evening, after they have worked all day, they will all eat out together or have a picnic in the penthouse. He has drawings of all this.

He has also been making fantasy sexy drawings of girls; Margaret discovered this by accident one day, while they were traveling, when she wandered into his hotel room and found a drawing by his bed. His other drawings—even the grandest ones, which he has spent days making—he is almost indifferent to; they can get lost or damaged, and he scarcely cares. But the sexy drawings are manifestly different; he seems to feel these as his own and keeps them in the privacy of his room—he would not think of showing them to anyone. They are wholly different from his other drawings, his commissioned work, for they are an expression of his inner life and dreams and needs, of his emotional and personal identity; whereas the architectural drawings, however dazzlingly accomplished, are not intended as anything more than likenesses, reproductions.

Stephen's interest in girls, his fantasies of them, all seem very normal, very adolescent in a way, and yet they are marked by a childishness, a naïveté that reflects his deep lack of human and social knowledge. It is difficult to imagine him dating, much less enjoying a deep personal or sexual relationship. These things, one suspects, may never be possible for him. I wonder whether he feels this, or feels sad about it sometimes.

In July of 1993, Margaret phoned me, beside herself with excitement. "Stephen's erupted musical powers," she announced. "*Huge* powers! You must come and see him straight-away." I was startled by her call; I had never known her so excited.

Stephen's musical talents clearly went back to early child-hood, like his artistic talents. Lorraine Cole writes that, even when he was scarcely verbal, he was a natural performer and mime: "His portrayal of an angry man in a restaurant was so spirited and so funny that it was only when we played back the video we had made that we realized he had used no actual words, only a wide range of angry noises. It was then that we understood his capacity for imitating sounds." This was espe-cially striking after a brief visit to Japan—the sound of the language fascinated him, and when Andrew picked him and Margaret up from Heathrow, Stephen babbled pseudo-Japanese, complete with "Japanese" gestures, to such effect that Andrew almost crashed the car laughing.

It had been clear to all of us, for years, that Stephen had an immense ability to reproduce instrumental sounds, voices, ac-cents, intonations, melodies, rhythms, arias, songs—complete with words or lyrics when need be—an effortlessly large and accurate auditory memory. And, significantly, he liked music, too; it moved him with an almost physical pleasure, almost more, I think, than drawing did.

But Margaret, who knew all this better than I, was obvi-ously referring to something more, to some quite new and un-expected breakthrough. The crucial factor, she had said, had been finding the right music teacher for Stephen ("She's mar-velous, darling!"), and they had struck up an instant rapport. I timed a visit to London to coincide with one of their weekly music lessons and took along my niece Liz Chase, a music teacher and pianist with a very acute ear, skilled in improvisa-tion, analysis, and theory.

Liz and I had been chatting with Evie Preston, his music teacher, for a few minutes when Stephen came in, gustily, at the stroke of twelve. "Hullo, Evie, how are you I am fine," he said, then, "Hullo Oliver Sacks, how are you?" and, when I in-troduced my niece, "Hullo Liz Chase, how are you?" He then rushed over to the piano and, under Evie's bidding, started to play scales, then to sing chords, starting with major triads. He did all this very easily, and gleefully. The idea of thirds, fifths—this Pythagorean, numerical sense of musical intervals—seemed quite innate in Stephen. "I never had to teach him," Evie remarked.

He seemed hungry for more. "Let's do sevenths now," Evie said, and Stephen nodded and chortled as if he had been promised a chocolate.

Next, Evie said, "Now we'll do the blues—you take the top, I'll do the bass." Using only three fingers (it looked ungainly, but worked brilliantly), Stephen now improvised an upper voice, full of intriguing, delightful complications. At first he confined his improvisations to the lower half of one octave, but then became bolder, his improvisations steadily becoming wider ranging, more complex. He did six improvisations in all, rising to a climax in the last one. But, Liz said, "Improvisation is easy, you do it off the top of your head." If one had the musical intelligence to catch the variational structure, she added, an ability to generate variations was almost automatic, a defining quality of intelligence itself. What she did find remarkable was how Stephen had infused his improvisations with feeling, with something of himself; how he had made them "creative, daring, and dramatically interesting."

Evie asked Stephen if he would sing "What a Wonderful World." His singing seemed to be full of genuine feeling, and his gestures while he sang were not his usual stilted, ticlike ones. As soon as the song was over, Evie asked Stephen to analyze it harmonically, to sing and number all the chords. He did so without a moment's hesitation. "It is clear that he is possessed of quite extraordinary powers of harmonic identification, analysis, and reproduction," Liz noted. Then Evie gave him an exercise in "interpretation," as she does every week, playing a theme he had never heard before, Schumann's "Träumerei." Stephen listened intently and told us his "associations" as he listened: "It's about . . . air in the field, daffo-

dils in springtime . . . a stream . . . sunshine . . . (I love it) . . .
rose gardens . . . light breezes, fresh . . . children come out to
play with their friends."

Was Stephen—so lacking in feeling or cut off from it, for the
most part—actually feeling these affects and moods? Or had
he learned, been taught somehow, to "decode" music, to learn
that such-and-such forms were "pastoral" or "vernal," and as
such would have appropriate images? Was this a sort of trick,
performed without any real feeling? I mentioned this thought
to Evie later, and she told me that at first his associations to
music were random or egocentric, strikingly irrelevant to the
actual tone of the piece. She then explained what feelings or
images "went with" different forms of music, and now he has
learned these. But she thinks he also feels them.

Finally, it was time for Stephen to choose a song he wanted
to perform. He wanted to do "It's Not Unusual," a song much
to his liking—a piece on which he could really let himself go.
He sang with great enthusiasm, swinging his hips, dancing,
gesticulating, miming, clutching an imaginary microphone to
his mouth, addressing himself in imagination to a vast arena.
"It's Not Unusual" has become the theme song of Tom Jones,
and in his version, Stephen took on Jones's flamboyant phys-
icality, adding to it a flavor of Stevie Wonder. He seemed completely at one with the music, completely possessed—and at
this point there was none of the skewed neck posture that is
habitual with him, none of the stiltedness, the ticcing, the
aversion of gaze. His entire autistic persona, it seemed, had totally vanished, replaced by movements that were free, grace-
ful, with emotional appropriateness and range. Very startled at
this transformation, I wrote in large capitals in my notebook,
"AUTISM DISAPPEARS." But as soon as the music stopped,
Stephen looked autistic once again.

Until now, it had seemed to be part of Stephen's nature, part
of being autistic, to be defective precisely in that range of
emotions and states of mind that defines a "self" for the rest
of us. And yet in the music he seemed to have been "given"

these, to have "borrowed" an identity—though these were lost the moment the music ended.

It was as if, for a brief time, he had become truly alive.

Stephen's music lesson, then, was a revelation to me—not just of further talents (not wholly unexpected in an autistic savant), but of a *mode of being* that I would not have thought available to him. Nothing of what I had seen with him before, and nothing in his art, had quite prepared me for this. He seemed to be using his whole self, his whole body, with all its repertoire of movements and expressions, to sing, to enact the song—though it remained unclear to me whether this was basically a brilliant piece of pantomime or a true entering into the words, the feelings, the inner states of the song. It raised for me (even more acutely than some of his Matisse drawings) the question of whether he treated the originals (paintings or songs) as representations of inwardness, of others' states of mind, or as *objects*. Did he, so to speak, enter the painter's or the songwriter's head, share their subjectivity, or merely treat their productions (like houses) as purely physical, as objects? (Was his repetition of *Rain Man*, for that matter, just a literal playback, a mimicry or echolalia, or was it charged with a sense of the significance of the film?) Were his gifts no more than mindless, "ament talents," in Goldstein's term, or were they genuine achievements of mind and identity?

Goldstein is quick to equate "mind" with the abstract-categorical, the conceptual, and to regard anything else as pathological, as sterile. But there are forms of health, of mind, other than the conceptual, although neurologists and psychologists rarely give these their due. There is mimesis—itself a power of mind, a way of representing reality with one's body and senses, a uniquely human capacity no less important than symbol or language. Merlin Donald, in *Origins of the Modern Mind*, has speculated that mimetic powers of modeling, of inner representation, of a wholly nonverbal and nonconceptual type, may have been the dominant mode of cognition for a million years or more in our immediate predecessor, *Homo erectus*, before the advent of abstract thought and language in

Homo sapiens.[11] As I watched Stephen sing and mime, I wondered if one might not understand at least some aspects of autism and savantism in terms of the normal development, even hypertrophy, of mimesis-based brain systems, this ancient mode of cognition, coupled with a relative failure in the development of more modern, symbol-based ones. And yet, even if some analogies can be drawn here, they are very partial and must not mislead us. Stephen is neither an ament, nor a computer, nor a *Homo erectus*—all our models, all our terms, break down before him.

Stephen's development has been singular, qualitatively different, from the start. He constructs the universe in a different way—and his mode of cognition, his identity, his artistic gifts, go together. We do not know, finally, how Stephen thinks, how he constructs the world, how he is able to draw and sing. But we do know that though he may be lacking in the symbolic, the abstract, he has a sort of genius for concrete or mimetic representations, whether drawing a cathedral, a canyon, a flower, or enacting a scene, a drama, a song—a sort of genius for catching the formal features, the structural logic, the style, the "thisness" (though not necessarily the "meaning"), of whatever he portrays.

Creativity, as usually understood, entails not only a "what," a talent, but a "who"—strong personal characteristics, a strong identity, personal sensibility, a personal style, which flow into the talent, interfuse it, give it personal body and form. Creativity in this sense involves the power to originate,

---

[11] Jerome Bruner, who has studied cognitive growth in children so minutely, speaks of "enactive" representation as its first expression. The enactive, he emphasizes, though it is supplemented by subsequently developed forms of cognition or representation (which he terms the "ikonic" and "symbolic"), is not superseded by them, but remains throughout life a potent mode of expression, instantly available for use. So it is with Donald's mimetic stage—this did not go out with *Homo erectus*, but remains a perpetual and powerful part of our own "sapient" repertoire. All of us make frequent use of such nonverbal behaviors and communications, and they are supremely developed in mimes, in actors, in all performing artists, and in the deaf.

to break away from the existing ways of looking at things, to move freely in the realm of the imagination, to create and re-create worlds fully in one's mind—while supervising all this with a critical inner eye. Creativity has to do with inner life—with the flow of new ideas and strong feelings.

Creativity, in this sense, will probably never be possible for Stephen. But the catching of thisness, perceptual genius, is no small gift; it is quite as rare and precious as more intellectual gifts. I once referred to José as living not in a universe, but in what William James called a "multiverse," of innumerable, unconnected though intensely vivid particulars, and as experiencing the world (in Proust's term) as "a collection of moments"—vivid, isolated, with no before or after. I imagined José, who liked to draw animals and plants, as an illustrator for botanical works or herbals (indeed, I have since heard that an autistic artist is employed by the Royal Botanical Gardens at Kew).

Is autism necessary to, or an ingredient of, his art? Most autists are not artists, as most artists are not autists; but in the chance of their coming together (as in Stephen, or José), there must, I think, be an interaction between the two, so that the art takes on some of the strengths and weaknesses of autism, its remarkable capacity for minutely detailed reproduction and representation, but also its repetitiveness and stereotypy. But whether one can speak of a distinctive "autistic art," I am not sure.

Is Stephen, or his autism, changed by his art? Here, I think, the answer is no. I do not have the feeling that his art spreads or diffuses, in any sense, into his character, or alters the general tone of his mind. But this, perhaps, is not entirely surprising: there are many examples of artists who are great, even sublime, in their art, but whose personal lives are unremarkable, incoherent, or vile. (There are others, of course, whose lives match their art.)

Of those with classical autism, 50 percent are mute, never use speech; 95 percent lead very limited lives—Stephen, in a sense, has escaped from these statistics, in part through his art, in part by virtue of those who have stood so committedly behind him. Gifts and art, unrecognized, unsupported, are not enough: José is almost as gifted as Stephen but has never been recognized, never supported, and continues to languish on a back ward; whereas Stephen lives a varied and stimulating life—he travels, goes out drawing, and now attends art school. Margaret Hewson, Chris Marris, and others have played an essential part in supporting him and nurturing his gifts, making possible for him his present creative life. But his passivity remains extreme, and he will continue, I think, to need such personal support, as Blind Tom needed the support of Colonel Bethune.

Stephen's drawings may never develop, may never add up to a major opus, an expression of a deep feeling or theory or view of the world. And *he* may never develop, or enter the full estate, the grandeur and misery, of being human, of man.

But this is not to diminish him, or to call his gifts small. His limitations, paradoxically, can serve as strengths, too. His vision is valuable, it seems to me, precisely because it conveys a wonderfully direct, unconceptualized view of the world. Stephen may be limited, odd, idiosyncratic, autistic; but it is given him to achieve what few of us do, a significant representation and investigation of the world.

An Anthropologist
on Mars

I had just returned from a few days with Stephen Wiltshire in July. I had driven up to Massachusetts to visit another autistic artist, Jessy Park (whose mother describes her in a most beautiful and intelligent personal narrative, "The Siege"), and had seen her intensely colored, star-studded drawings (very different from Stephen's) and something of her labyrinthine, magic world of correlations (between numbers, colors, morality, the weather). I had paid flying visits to several schools for autistic children. I had spent an extraordinary week at a camp for autistic children, Camp Winston, in Ontario—the more so as one of the counselors there this summer was a friend of mine, Shane, with Tourette's syndrome, who, with his lungings and touchings, reachings and buttings, his enormous vitality and impulsiveness, seemed able to get through to the most deeply autistic children, in a way the rest of us were unable to do. Turning west, I had visited an entire autistic family in California—both parents, highly gifted, and their two children, all of them given (between the serious business of life) to jumping on trampolines, flapping their hands, and screaming. And now, finally, I was on my way to Fort Collins, in Colorado, to see Temple Grandin, one of the most remarkable autistic people of all: in spite of her autism, she holds a Ph.D. in animal science, teaches at Colorado State University, and runs her own business.

While autism was described almost simultaneously by Leo Kanner and Hans Asperger in the 1940s, Kanner seemed to see it as an unmitigated disaster, where Asperger felt that it might have certain positive or compensating features—a "particular originality of thought and experience, which may well lead to exceptional achievements in later life."

It is clear even in these first accounts that there is a wide range of phenomena and symptoms in autism—and many more can be added to those that Kanner and Asperger listed. A majority of Kanner-type children are retarded, often severely; a significant proportion have seizures and may have "soft" neurological signs and symptoms—a whole range of repetitive or automatic movements, such as spasms, tics, rocking, spinning, finger play, or flapping of the hands; problems of coordination and balance; peculiar difficulties, sometimes, in initiating movements, akin to what is seen in parkinsonism. There may also be, very prominently, a large range of abnormal (and often "paradoxical") sensory responses, with some sensations being heightened and even intolerable, others (which may include pain perception) being diminished or apparently absent. There may be, if language develops, odd and complex language disorders—a tendency to verbosity, empty chatter, cliché-ridden and formulaic speech; the psychologist Doris Allen describes this aspect of their autism as a "semantic-pragmatic deficit." In contrast, Asperger-type children are often of normal (and sometimes very superior) intelligence and generally have fewer neurological problems.

Kanner and Asperger looked at autism clinically, providing descriptions of such fullness and accuracy that even now, fifty years later, they can hardly be bettered. But it was not until the 1970s that Beate Hermelin and Neil O'Connor and their colleagues in London, trained in the new discipline of cognitive psychology, focused on the mental structure of autism in a more systematic way. Their work (and that of Lorna Wing, in particular) suggests that in all autistic individuals there is a core problem, a consistent triad of impairments: impairment of social interaction with others, impairment of verbal and nonverbal communication, and impairment of play and imaginative activities. The appearance of these three together, they feel, is not fortuitous; all are expressive of a single, fundamental developmental disturbance. Autistic people, they suggest, have no true concept of, or feeling for, other minds, or even their own; they have, in the jargon of cognitive psychology, no "theory of mind." However, this is only one hypothesis among many; no theory, as yet, encompasses the whole range of phenomena to be seen in autism. Kanner and Asperger were still, in the 1970s, pondering the syndromes they had delineated more than thirty years earlier, and the foremost workers of today have all spent twenty years or more considering them. Autism as a subject touches on the deepest questions of ontology, for it involves a radical deviation in the development of brain and mind. Our insight is advancing, but tantalizingly slowly. The ultimate understanding of autism may demand both technical advances and conceptual ones beyond anything we can now even dream of.

The picture of "classical infantile autism" is a formidable one. Most people (and, indeed, most physicians), if asked about autism, summon up a picture of a profoundly disabled child, with stereotyped movements, perhaps head-banging; rudimentary language; almost inaccessible: a creature for whom very little future lies in store.

Indeed, in a strange way, most people speak only of autistic children and never of autistic adults, as if the children somehow just vanished from the earth. But though there may indeed be a devastating picture at the age of three, some autistic youngsters, contrary to expectations, may go on to develop fair language, a modicum of social skills, and even high intellectual achievements; they may develop into autonomous human beings, capable of a life that may at least appear full and normal—even though, beneath it, there may remain a persistent, and even profound, autistic singularity. Asperger had a clearer idea of this possibility than Kanner; hence we now speak of such "high-functioning" autistic individuals as having Asperger's syndrome. The ultimate difference, perhaps, is this: people with Asperger's syndrome can tell us of their experiences, their inner feelings and states, whereas those with classical autism cannot. With classical autism, there is no window, and we can only infer. With Asperger's syndrome there is self-consciousness and at least some power to introspect and report.

Whether Asperger's syndrome is radically different from classical infantile autism (in a child of three, all forms of autism may look the same) or whether there is a continuum from the severest cases of infantile autism (accompanied, perhaps, by retardation and various neurological problems) to the most gifted, high-functioning individuals, is a matter of dispute. (Isabelle Rapin, a neurologist who specializes in autism, stresses that the two conditions may be separate at the biological level even if they are sometimes similar at the behavioral level.) It is also unclear whether this continuum should be extended to include the possession of isolated "autistic traits"—peculiar, intense preoccupations and fixations, often combined with relative social withdrawal or remoteness—such as one encounters in any number of people conventionally called "normal" or seen, at most, as a little odd, eccentric, pedantic, or reclusive.

The cause of autism has also been a matter of dispute. Its incidence is about one in a thousand, and it occurs throughout the world, its features remarkably consistent even in extremely different cultures. It is often not recognized in the first year of life, but tends to become obvious in the second or third year. Though Asperger regarded it as a biological defect of affective contact—innate, inborn, analogous to a physical or intellectual defect—Kanner tended to view it as a psychogenic disorder, a reflection of bad parenting, and most especially of a chillingly remote, often professional, "refrigerator mother." At this time, autism was often regarded as "defensive" in nature, or confused with childhood schizophrenia. A whole generation of parents—mothers, particularly—were made to feel guilty for the autism of their children. It was only in the 1960s that this trend began to reverse, and the organic nature of autism to be fully accepted. (Bernard Rimland's 1964 text, *Infantile Autism*, played an important part here.)

That the disposition to autism is biological is no longer in doubt, nor the increasing evidence that it is, in some cases, genetic. Genetically, autism is heterogeneous—it is sometimes dominant, sometimes recessive. It is much more common in males. The genetic form may be associated, in the affected individual or the family, with other genetic disorders, such as dyslexia, attention deficit disorder, obsessive-compulsive disorder, or Tourette's syndrome. But autism may also be acquired. This was first realized in the 1960s with the epidemic of rubella, when a large number of babies exposed to this prenatally went on to develop autism. It remains unclear whether the so-called regressive forms of autism—with sometimes abrupt losses of language and social behavior in two- to four-year-olds who had previously been developing relatively normally—are genetically or environmentally caused. Autism may be a consequence of metabolic problems (such as phenylketonuria) or mechanical ones (such as hydrocephalus).[1] Autism, or autismlike syndromes, may develop even in adult life, though infrequently, especially after certain forms of encephalitis. (Some of my *Awakenings* patients, I think, had elements of autism, too.)

And yet the parents of an autistic child, who find their infant receding from them, becoming remote, inaccessible, unresponsive, may still be tempted to blame themselves. They may find themselves struggling to relate to and love a child who, seemingly, does not love them back. They may make superhuman efforts to get through, to hold on to a child who in-

---

[1] The television show "20/20" has reported on a town in Massachusetts with a very high incidence of autism, especially in the neighborhood of a former plastics factory—but the question of whether autism can be caused by exposure to toxic agents has yet to be fully studied.

habits some unimaginable, alien world; and yet all their efforts may seem to be in vain.

The history of autism, indeed, has been in part a desperate search for, and promotion of, "breakthroughs" of various sorts. One father of an autistic boy expressed this to me with some bitterness: "They come up with a new 'miracle' every four years—first it was elimination diets, then magnesium and vitamin B_6, then forced holding, then operant conditioning and behavior modification—now all the excitement is about auditory desensitization and facilitated communication." This boy, at twelve, was still tantalizingly mute and unreachable, and his condition had defied every form of attempted therapy—hence his father's pessimism and blanket condemnation. Responses seem to be extremely varied: some individuals may respond spectacularly to some of these methods, while others show virtually no response at all.[2]

---

[2] The most recent and controversial of these methods is facilitated communication. FC (originally used with children with cerebral palsy) is based on the notion that if the hand or arm of a nonverbal autistic child is supported by a facilitator, the child may then be able to communicate by typing or by using an electronic communicator or letter board. The underlying thought is that such children may have a difficulty in initiating movements (akin to that of parkinsonism), and that a light contact with another person may allow them to overcome this and achieve a normal motor facility (as may occur with touching, or even visual contact, in some parkinsonian patients—I discuss this in *Awakenings*, footnote 45). The hope is that there may be, in at least some otherwise inaccessible patients, a rich but "imprisoned" world of thought and feeling that may now be released by this simple tactic.

The reported range of effects is very great, from minor releases of simple communications in some patients to entire autobiographies seemingly emanating from previously mute children. These reports have been the subject of almost evangelistic enthusiasm, among many parents and teachers of autistic children on the one hand; and of wholesale dismissal by the medical profession, on the other. It has been difficult to arrive at a calm judgment in the overcharged atmosphere of claims and dismissals; while some instances of FC have been shown to be entirely factitious—the result of unconscious suggestion by the facilitator—and others must be suspect, there remains a nucleus of apparently bona fide phenomena that deserve a careful and openminded scrutiny.

No two people with autism are the same; its precise form or expression is different in every case. Moreover, there may be a most intricate (and potentially creative) interaction between the autistic traits and the other qualities of the individual. So, while a single glance may suffice for clinical diagnosis, if we hope to understand the autistic individual, nothing less than a total biography will do.

My own first experience with the autistic was in a grim ward in a state hospital in the midsixties. Many of these patients, perhaps a majority, were also retarded; many had seizures; many had violent self-abusive behaviors, such as head-banging; many had other neurological problems. These worst-off patients tended to be multiply handicapped in addition to their autism (and several had been traumatized by abuse). And yet, even in this population, there were sometimes "islands of ability," occasionally spectacular talents, shining through the devastation, precisely as Kanner and Asperger had described—remarkable numerical or graphic powers, for instance. It was these special talents, apparently isolated from the rest of the mind and personality, and maintained by a passionate, intensely focused fixation or motivation—these savant syndromes—that engaged my special interest and that I explored most deeply at the time. And even in this population of the seemingly hopeless, there were some who responded to individual attention. One young patient, nonverbal, responded to music and danced; another, after some weeks, started to play pool with me and later, in the botanical garden, said his first word—"dandelion." Many of these patients, born in the 1940s or early 1950s, had not even been diagnosed as autistic when young, but had been lumped together indiscriminately with the retarded and psychotic and warehoused in huge institutions since early childhood. This is probably how the severely autistic have been treated for centuries. It has only been in the last two decades or so that the picture for such youngsters has decisively changed, with increasing medical and educational awareness of their special strengths and problems, and the widespread introduction of special schools and camps for autistic children.[3]

Visiting a few of these during August, I had seen a variety of children, some intelligent, some mildly retarded, some outgoing, some timid, all with their own individual personalities. At one such school, as I approached, I had seen some children in the playground, swinging and playing ball. How normal, I thought—but when I got closer I saw one child swinging obsessively in terrifying semicircles, as high as the swing would go; another throwing a small ball monotonously from hand to hand; another spinning on a roundabout, around and around; another not building with bricks but lining them up endlessly, in neat, monotonous rows. All were engaged in solitary, repetitive activities; none was really playing, or playing with any of the others. Some of the children inside, when not in classes, would rock back and forth; some would flap their hands or jabber unintelligibly. Occasionally, one of the teachers told me, a few of the children would have sudden panics or rages and scream or hit out uncontrollably. Some of the children would echo any words that were spoken to them. One boy apparently had an entire television show by heart and would "replay" it all day, complete with all the voices and gestures, and even sounds of applause. At Camp Winston, an attractive six-year-old boy had been given a pair of scissors and was cutting minute "H"s, a fraction of an inch high, each perfect, from a piece of paper. Most of the children looked physically normal—it was their remoteness, their inaccessibility, that were so uncanny.

Some, in adolescence, were starting to emerge—to speak fluently, to learn social skills (much more difficult for such children than any academic learning), to create social surfaces they could present to the world.

Without special schooling—schooling that for many had

---

[3] A pioneer here was Mira Rothenberg, who formed the Blueberry Treatment Centers in 1958, an early experience she describes in her book, *Children with Emerald Eyes.*

started in the nursery or at home—these autistic youngsters, despite their often good intelligence and background, might have remained profoundly isolated and disabled. They had certainly learned, many of them, to "operate" after a fashion, to show at least a formal or external recognition of social conventions—and yet the very formality or externality of their behavior was itself disconcerting. I felt this especially at one school I visited, where children would stick out rigid hands and say in loud, unmodulated voices, "Good morning my name is Peter . . . I am very well thank you how are you" without any punctuation or intonation, affect or tone, in a sort of litany. Would any of them, I wondered, ever achieve true autonomy? Use their social automatisms pragmatically, as a way of functioning in the world, but, beyond this, achieve a true inwardness of their own, perhaps a profoundly different inner life, of an autistic sort—perhaps an inner life known or shown only to a few others?

Uta Frith has written, in her book *Autism: Explaining the Enigma*, "Autism . . . does not go away. . . . Nevertheless, autistic people can, and often do, compensate for their handicap to a remarkable degree. [But] there remains a persistent deficit . . . something that cannot be corrected or substituted." She also implies, in a speculative mood, that there may be a reverse side to this "something," a sort of moral or intellectual intensity or purity, so far removed from the normal as to seem noble, ridiculous, or fearful to the rest of us. She wonders, in this regard, about the blessed fools of old Russia, about the ingenuous Brother Juniper, an early follower of Saint Francis, and, interestingly, about Sherlock Holmes, with his oddness, his peculiar fixations—his "little monograph on the ashes of 140 different varieties of pipe, cigar and cigarette tobacco," his "clear powers of observation and deduction, unclouded by the everyday emotions of ordinary people," and the extreme unconventionality that often allows him to solve a case that the police, with their more conventional minds, are unable to solve. Asperger himself wrote of "autistic intelligence" and saw it as a sort of intelligence scarcely touched by tradition and culture—unconventional, unorthodox, strangely "pure" and original, akin to the intelligence of true creativity.

Dr. Frith, when we met in London, expanded on these themes and said I must be sure to visit one of the most remarkable autistic people she knew—to see her at work and at home, to spend time with her. "Go see Temple," Dr. Frith said as I left her office.

I had, of course, heard of Temple Grandin—everyone interested in autism has heard of her—and had read her autobiography, *Emergence: Labeled Autistic*, when it came out, in 1986. When I first read the book, I could not help being suspicious of it: the autistic mind, it was supposed at that time, was incapable of self-understanding and understanding others and therefore of authentic introspection and retrospection. How *could* an autistic person write an autobiography? It seemed a contradiction in terms. When I observed that the book had been written in collaboration with a journalist, I wondered whether some of its fine and unexpected qualities—its coherence, its poignancy, its often "normal" tone—might in fact be due to her. Such suspicions have continued to be voiced, in regard to Grandin's book and to autistic autobiographies in general, but as I read Temple's papers (and her many autobiographical articles) I found a detail and consistency, a directness, that changed my mind.[4]

Reading her autobiography and her articles, one gets a feeling of how strange, how different, she was as a child, how far

---

[4] What one does see in Temple's writings (and in the writings of other very able autistic adults, not excluding some with marked literary gifts) are peculiar narrational gaps and discontinuities, sudden, perplexing changes of topic, brought about (so Francesca Happé suggests in a recent essay on the subject) by Temple's failure "to appreciate that her reader does not share the important background information that she possesses." In more general terms, autistic writers seem to get "out of tune" with their readers, fail to realize their own or their readers' states of mind.

removed from normal.[5] At six months, she started to stiffen in her mother's arms, at ten months to claw her "like a trapped animal." Normal contact was almost impossible in these circumstances. Temple describes her world as one of sensations heightened, sometimes to an excruciating degree (and inhibited, sometimes to annihilation): she speaks of her ears, at the age of two or three, as helpless microphones, transmitting everything, irrespective of relevance, at full, overwhelming volume—and there was an equal lack of modulation in all her senses. She showed an intense interest in odors and a remarkable sense of smell. She was subject to sudden impulses and, when these were frustrated, violent rage. She perceived none of the usual rules and codes of human relationship. She lived, sometimes raged, inconceivably disorganized, in a world of unbridled chaos. In her third year, she became destructive and violent:

> Normal children use clay for modelling; I used my feces and then spread my creations all over the room. I chewed up puzzles and spit the cardboard mush out on the floor. I had a violent temper, and when thwarted, I'd throw anything handy—a museum quality vase or leftover feces. I screamed continually . . .

And yet, like many autistic children, she soon developed an immense power of concentration, a selectivity of attention so intense that it could create a world of its own, a place of calm and order in the chaos and tumult: "I could sit on the beach for hours dribbling sand through my fingers and fashioning miniature mountains," she writes. "Each particle of sand intrigued me as though I were a scientist looking through a microscope. Other times I scrutinized each line in my finger,

---

[5] Authentic memories from the second (perhaps even the first) year of life, though not available to "normals," may be recalled, with veridical detail, by autistic people. Thus, Lucci et al. write of one such boy, "He seems to recall, in exquisite detail, events from when he was two or three years old." Coenesthetic memories of infancy are also reported by Luria of S., the mnemonist he studied.

following one as if it were a road on a map." Or she would spin, or spin a coin, so raptly that she saw and heard nothing else. "People around me were transparent. . . . Even a sudden loud noise didn't startle me from my world." (It is not clear whether this hyperfocus of attention—an attention as narrow as it is intense—is a primary phenomenon in autism or a reaction or adaptation to an overwhelming, uninhibited barrage of sensation. A similar hyperfocus is sometimes seen in Tourette's syndrome.)

At three, Temple was taken to a neurologist, and the diagnosis of autism was made; it was hinted that lifelong institutionalization would probably be necessary. The total absence of speech at this age seemed especially ominous.

How, I had to wonder, had she ever moved from this almost unintelligible childhood, with its chaos, its fixations, its inaccessibility, its violence—this fierce and desperate state, which had almost led to her institutionalization at the age of three—to the successful biologist and engineer I was going to see?

I phoned Temple from the Denver airport to reconfirm our meeting—it was conceivable, I thought, that she might be somewhat inflexible about arrangements, so time and place should be set as definitely as possible. It was an hour-and-a-quarter drive to Fort Collins, Temple said, and she provided minute directions for finding her office at Colorado State University, where she is an assistant professor in the Animal Sciences Department. At one point, I missed a detail, and asked Temple to repeat it, and was startled when she repeated the entire directional litany—several minutes' worth—in virtually the same words. It seemed as if the directions had to be given as they were held in Temple's mind, entire—that they had fused into a fixed association or program and could no longer be separated into their components. One instruction, however, had to be modified. She had told me at first that I should turn right onto College Street at a particular intersection marked by a Taco Bell restaurant. In her second set of directions, Tem-

ple added an aside here, said the Taco Bell had recently had a
face-lift and been housed in a fake cottage, and no longer
looked in the least "bellish." I was struck by the charming,
whimsical adjective "bellish"—autistic people are often called
humorless, unimaginative, and "bellish" was surely an original concoction, a spontaneous and delightful image.

I made my way to the university campus and located the
Animal Sciences Building, where Temple was waiting to greet
me. She is a tall, strongly built woman in her midforties; she
was wearing jeans, a knit shirt, western boots, her habitual
dress. Her clothing, her appearance, her manner, were plain,
frank, and forthright; I had the impression of a sturdy, no-
nonsense cattlewoman, with an indifference to social conventions, appearance, or ornament, an absence of frills, an
absolute directness of manner and mind. When she raised her
arm in greeting, the arm went too high, seemed to get caught
for a moment in a sort of spasm or fixed posture—a hint, an
echo, of the stereotypies she once had. Then she gave me a
strong handshake and led the way down to her office. (Her gait
seemed to me slightly clumsy or uncouth, as is often the case
with autistic adults. Temple attributes this to a simple ataxia
associated with impaired development of the vestibular system and part of the cerebellum. Later I did a brief neurological
exam, focusing on her cerebellar function and balance; I did
indeed find a little ataxia, but insufficient, I thought, to explain her odd gait.)

She sat me down with little ceremony, no preliminaries, no
social niceties, no small talk about my trip or how I liked Colorado. Her office, crowded with papers, with work done and to
do, could have been that of any academic, with photographs of
her projects on the wall and animal knickknacks she had
picked up on her travels. She plunged straight into talking of
her work, speaking of her early interests in psychology and animal behavior, how they were connected with self-observation
and a sense of her own needs as an autistic person, and how
this had joined with the visualizing and engineering part of
her mind to point her toward the special field she had made

her own: the design of farms, feedlots, corrals, slaughter-houses—systems of many sorts for animal management.

She handed me a book containing some of the layouts she had developed over the years—the book was titled *Beef Cattle Behaviors, Handling, and Facilities Design*—and I admired the complex and beautiful designs inside, and the logical presentation of the book, starting with diagrams of cattle and sheep and hog behavior and moving through designs of corrals to ever more complex ranch and feedlot facilities.

She spoke well and clearly, but with a certain unstoppable impetus and fixity. A sentence, a paragraph, once started, had to be completed; nothing was left implicit, hanging in the air.

I was feeling somewhat exhausted, hungry, and thirsty—I had been traveling all day and had missed lunch—and I kept hoping Temple would notice and offer me some coffee. She did not; so, after an hour, almost fainting under the barrage of her overexplicit and relentless sentences, and the need to attend to several things at once (not only what she was saying, which was often complex and unfamiliar, but also her mental processes, the sort of person she was), I finally asked for some coffee. There was no "I'm sorry, I should have offered you some before," no intermediacy, no social junction. Instead, she immediately took me to a coffeepot that was kept brewing in the secretaries' office upstairs. She introduced me to the secretaries in a somewhat brusque manner, giving me the feeling, once again, of someone who had learned, roughly, "how to behave" in such situations without having much personal perception of how other people felt—the nuances, the social subtleties, involved.

Time to get some dinner," Temple suddenly announced after we had spent another hour in her office. "We eat early in the West." We went to a nearby western restaurant, one with swinging doors and with guns and cattle horns on the walls—it was already crowded, as Temple had said it would be, at five in the afternoon—and we ordered a classic western meal of ribs and beer. We ate heartily and talked throughout

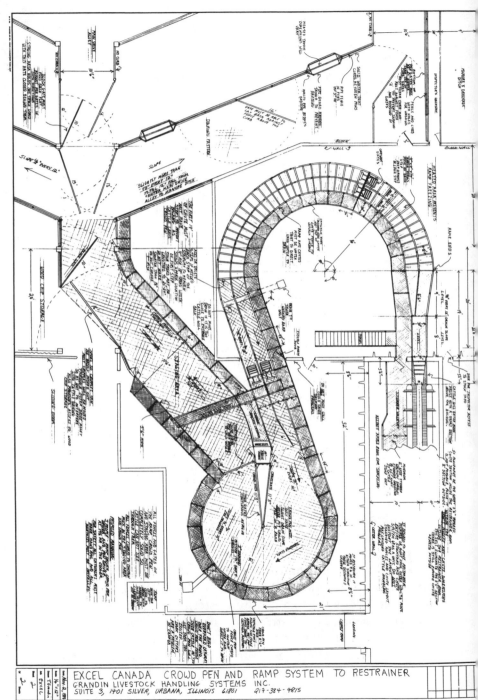

EXCEL CANADA CROWD PEN AND RAMP SYSTEM TO RESTRAINER
GRANDIN LIVESTOCK HANDLING SYSTEMS INC.
SUITE 3, 1401 SILVER, URBANA, ILLINOIS 61801   217-384-4815

the meal about the technical aspects of Temple's work and the ways in which she sets out every design, every problem, visually, in her mind. As we left the restaurant, I suggested we go for a walk, and Temple took me out to a meadow along an old railway line. The day was cooling rapidly—we were at five thousand feet—and in the long evening light gnats darned the air and crickets were stridulating all around us. I found some horsetails (one of my favorite plants) in a muddy patch below the tracks and became excited about them. Temple glanced at them, said "Equisetum," but did not seem stirred by them, as I was.

On the plane to Denver, I had been reading a remarkable piece of writing by a highly gifted, normal nine-year-old—a fairy story she had created, with a wonderful sense of myth, a whole world of magic, animism, and cosmogonies. What, I wondered as we walked through the horsetails, of Temple's cosmogony? How did she respond to myths, or to dramas? How much did they carry meaning for her? I asked her about the Greek myths. She said that she had read many of them as a child, and that she thought of Icarus in particular—how he had flown too near the sun and his wings had melted and he had plummeted to his death. "I understand Nemesis and Hubris," she said. But the loves of the gods, I ascertained, left her unmoved—and puzzled. It was similar with Shakespeare's plays. She was bewildered, she said, by Romeo and Juliet ("I never knew what they were up to"), and with *Hamlet* she got lost with the back-and-forth of the play. Though she ascribed these problems to "sequencing difficulties," they seemed to arise from her failure to empathize with the characters, to follow the intricate play of motive and intention. She said that she could understand "simple, strong, universal" emotions but was stumped by more complex emotions and the games people play. "Much of the time," she said, "I feel like an anthropologist on Mars."

She was at pains to keep her own life simple, she said, and to make everything very clear and explicit. She had built up a vast library of experiences over the years, she went on. They were like a library of videotapes, which she could play in her mind and inspect at any time—"videos" of how people behaved in different circumstances. She would play these over and over again and learn, by degrees, to correlate what she saw, so that she could then predict how people in similar circumstances might act. She had complemented her experience by constant reading, including reading of trade journals and the *Wall Street Journal*—all of which enlarged her knowledge of the species. "It is strictly a logical process," she explained.

In one plant she had designed, she said, there had been repeated breakdowns of the machinery, but these occurred only when a particular man, John, was in the room. She "correlated" these incidents and inferred at last that John must be sabotaging the equipment. "I had to learn to be suspicious, I had to learn it cognitively. I could put two and two together, but I couldn't see the jealous look on his face." Such incidents have not been uncommon in her life: "It bends some people out of shape that this autistic weirdo can come in and design all the equipment. They want the equipment, but it galls them that they can't do it themselves, but that Tom"—an engineering colleague—"and I can, that we've got hundred-thousand-dollar Sun workstations in our heads." In her ingenuousness and gullibility, Temple was at first a target for all sorts of tricks and exploitations; this sort of innocence or guilelessness, arising not from moral virtue but from failure to understand dissembling and pretense ("the dirty devices of the world," in Traherne's phrase), is almost universal among the autistic. But over the years Temple has learned, in her indirect way, by inspecting her "library," some of the ways of the world. She has, in fact, been able to found her own company and to work as a freelance consultant to and designer of animal facilities all over the world. By professional standards, she is extraordinarily successful, but other human interactions—social, sexual—she cannot "get." "My work is my life," she told me several times. "There is not that much else."

There seemed to me pain, renunciation, resolution, and acceptance all mixed together in her voice, and these are the feelings that sound through her writings. In one article she writes:

I do not fit in with the social life of my town or university. Almost all of my social contacts are with livestock people or people interested in autism. Most of my Friday and Saturday nights are spent writing papers and drawing. My interests are factual and my recreational reading consists mostly of science and livestock publications. I have little interest in novels with complicated interpersonal relationships, because I am unable to remember the sequence of events. Detailed descriptions of new technologies in science fiction or descriptions of exotic places are much more interesting. My life would be horrible if I did not have my challenging career.

Early the next morning, a Saturday, Temple picked me up in her four-wheel-drive, a rugged vehicle she drives all over the West to visit farms, ranches, corrals, and meat plants. As we headed for her house, I quizzed her about the work she had done for her Ph.D.; her thesis was on the effects of enriched and impoverished environments on the development of pigs' brains. She told me about the great differences that developed between the two groups—how sociable and delightful the "enriched" pigs became, how hyperexcitable and aggressive (and almost "autistic") the "impoverished" ones were by contrast. (She wondered whether impoverishment of experience was not a contributing factor in human autism.) "I got to love my enriched pigs," she said. "I was very attached. I was so attached I couldn't kill them." The animals had to be sacrificed at the end of the experiment so their brains could be examined. She described how the pigs, at the end, trusting her, let her lead them on their last walk, and how she had calmed them, by stroking them and talking to them, while they were killed. She was very distressed at their deaths—"I wept and wept."

She had just finished the story when we arrived at her home—a small two-story town house, some distance from the campus. Downstairs was comfortable, with the usual amenities—a sofa, armchairs, a television, pictures on the wall—but I had the sense that it was rarely used. There was an immense sepia print of her grandfather's farm in Grandin, North Dakota, in 1880; her other grandfather, she told me, had invented the automatic pilot for planes. These two were the progenitors, she feels, of her agricultural and engineering talents. Upstairs was her study, with her typewriter (but no word processor), absolutely bursting with manuscripts and books— books everywhere, spilling out of the study into every room in the house. (My own little house was once described as "a machine for working," and I had a somewhat similar impression of Temple's.) On one wall was a large cowhide with a huge collection of identity badges and caps, from the hundreds of conferences she has lectured at. I was amused to see, side by side, an I.D. from the American Meat Institute and one from the American Psychiatric Association. Temple has published more than a hundred papers, divided between those on animal behavior and facilities management and those on autism. The intimate blending of the two was epitomized by the medley of badges side by side.

Finally, without diffidence or embarrassment (emotions unknown to her), Temple showed me her bedroom, an austere room with whitewashed walls and a single bed and, next to the bed, a very large, strange-looking object. "What is that?" I asked.

"That's my squeeze machine," Temple replied. "Some people call it my hug machine."

The device had two heavy, slanting wooden sides, perhaps four by three feet each, pleasantly upholstered with a thick, soft padding. They were joined by hinges to a long, narrow bottom board to create a V-shaped, body-sized trough. There was a complex control box at one end, with heavy-duty tubes leading off to another device, in a closet. Temple showed me this as well. "It's an industrial compressor," she said, "the kind they use for filling tires."

"And what does this do?"

"It exerts a firm but comfortable pressure on the body, from the shoulders to the knees," Temple said. "Either a steady pressure or a variable one or a pulsating one, as you wish," she added. "You crawl into it—I'll show you—and turn the compressor on, and you have all the controls in your hand, here, right in front of you."

When I asked her why one should seek to submit oneself to such pressure, she told me. When she was a little girl, she said, she had longed to be hugged but had at the same time been terrified of all contact. When she was hugged, especially by a favorite (but vast) aunt, she felt overwhelmed, overcome by sensation; she had a sense of peacefulness and pleasure, but also of terror and engulfment. She started to have daydreams—she was just five at the time—of a magic machine that could squeeze her powerfully but gently, in a huglike way, and in a way entirely commanded and controlled by her. Years later, as an adolescent, she had seen a picture of a squeeze chute designed to hold or restrain calves and realized that that was it: a little modification to make it suitable for human use, and it could be her magic machine. She had considered other devices—inflatable suits, which could exert an even pressure all over the body—but the squeeze chute, in its simplicity, was quite irresistible.

Being of a practical turn of mind, she soon made her fantasy come true. The early models were crude, with some snags and glitches, but she eventually evolved a totally comfortable, predictable system, capable of administering a "hug" with whatever parameters she desired. Her squeeze machine had worked exactly as she hoped, yielding the very sense of calmness and pleasure she had dreamed of since childhood. She could not have gone through the stormy days of college without her squeeze machine, she said. She could not turn to human beings for solace and comfort, but she could always turn to it. The machine, which she neither exhibited nor concealed but kept openly in her room at college, excited derision and suspicion and was seen by psychiatrists as a "regression" or "fixation"—something that needed to be psychoanalyzed and resolved. With her characteristic stubbornness, tenacity, single-mindedness, and bravery—along with a complete absence of inhibition or hesitation—Temple ignored all these comments and reactions and determined to find a scientific "validation" of her feelings.

Both before and after writing her doctoral thesis, she made a systematic investigation of the effects of deep pressure in autistic people, college students, and animals, and recently a paper of hers on this was published in the *Journal of Child and Adolescent Psychopharmacology*. Today, her squeeze machine, variously modified, is receiving extensive clinical trials. She has also become the world's foremost designer of squeeze chutes for cattle and has published, in the meat-industry and veterinary literature, many articles on the theory and practice of humane restraint and gentle holding.

While telling me this, Temple knelt down, then eased herself, facedown and at full length, into the "V," turned on the compressor (it took a minute for the master cylinder to fill), and twisted the controls. The sides converged, clasping her firmly, and then, as she made a small adjustment, relaxed their grip slightly. It was the most bizarre thing I had ever seen, and yet, for all its oddness, it was moving and simple. Certainly there was no doubt of its effect. Temple's voice, often loud and hard, became softer and gentler as she lay in her machine. "I concentrate on how gently I can do it," she said, and then spoke of the necessity of "totally giving in to it. . . . I'm getting real relaxed now," she added quietly. "I guess others get this through relation with other people."

It is not just pleasure or relaxation that Temple gets from the machine but, she maintains, a feeling for others. As she lies in her machine, she says, her thoughts often turn to her mother, her favorite aunt, her teachers. She feels their love for her, and hers for them. She feels that the machine opens a door into an otherwise closed emotional world and allows her, almost teaches her, to feel empathy for others.

After twenty minutes or so, she emerged, visibly calmer, emotionally less rigid (she says that a cat can easily sense the difference in her at these times), and asked me if I would care to try the machine.

Indeed, I was curious and scrambled into it, feeling a little foolish and self-conscious—but less so than I might have been, because Temple herself was so wholly lacking in self-consciousness. She turned the compressor on again and filled the master cylinder, and I experimented gingerly with the controls. It was indeed a sweet, calming feeling—one that reminded me of my deep-diving days long ago, when I felt the pressure of the water on my diving suit as a whole-body embrace.

After my own trial in the squeeze machine, and with both of us suitably relaxed, we drove out to the university's experimental farm, where Temple does much of her basic fieldwork. I had earlier thought there might be a separation, even a gulf, between the personal—and, so to speak, private—realm of her autism and the public realm of her professional expertise. But it was becoming increasingly clear to me that they were hardly separated at all; for her, the personal and the professional, the inward and the outward, were completely fused.

"Cattle are disturbed by the same sorts of sounds as autistic people—high-pitched sounds, air hissing, or sudden loud noises; they cannot adapt to these," Temple told me. "But they are not bothered by low-pitched, rumbling noises. They are disturbed by high visual contrasts, shadows or sudden movements. A light touch will make them pull away, a firm touch calms them. The way I would pull away from being touched is the way a wild cow will pull away—getting me used to being touched is very similar to taming a wild cow." It was precisely her sense of the common ground (in terms of basic sensations and feelings) between animals and people that allowed her to show such sensitivity to animals, and to insist so forcefully on their humane management.

She had been primed to this knowledge, she felt, partly through the experience of her own autism and partly because she came from a long line of farmers and, as a child, had spent much of her time on farms. And her own mode of thinking allowed her no escape from these realities. "If you're a visual thinker, it's easier to identify with animals," she said as we drove to the farm. "If all your thought processes are in language, how could you imagine that cattle think? But if you think in pictures . . ."

Temple has always been a powerful visualizer. She was astonished when she discovered that her own near-hallucinatory power of visual imagery was not universal—that there were others who, apparently, had other ways to think. She is still very puzzled by this. "How *do* you think?" she kept asking me. But she had no sense that she could draw, make blueprints, until she was twenty-eight, when she met a draftsman and watched him drawing plans. "I saw how he did it," she told me. "I went and got exactly the same instruments and pencils as he used—a point-five-millimeter HB Pentel—and then I started pretending I was him. The drawing did itself, and when it was all done I couldn't believe I'd done it. I didn't have to learn how to draw or design, I pretended I was David—I appropriated him, drawing and all."[6]

---

[6] At first it seemed, from what Temple told me, that the "appropriated" David, and his skill, had been swallowed whole, existed only as a sort of implant or foreign body within her and was only slowly integrated to become part of her. Another gifted (and poetic) autistic woman has compared herself, in this regard, to a boa constrictor, swallowing entire animals whole, but only very slowly being able to assimilate them. Sometimes the swallowed role or skill seems not to be properly assimilated or integrated and may be lost or expelled as suddenly as it was acquired—thus the tendency (especially marked in younger autistic savants) to engulf complex skills or personas or masses of information wholesale, to juggle with these for a while, and then suddenly to relinquish or forget them with such completeness that they seem to pass through without leaving any residue whatever (such unincorporated behaviors and convulsive mimeses are sometimes seen in people with severe Tourette's syndrome).

Much more complex are the situations where behaviors, and indeed entire personas, are retained as a sort of pseudopersonality. The taking on of exaggerated, stereotypic, almost cartoonlike sexual demeanors (mimicked or caricatured from comic strips or soap operas on TV) is sometimes seen in adolescents with autism. Donna Williams, in her fascinating personal narratives (*Nobody Nowhere* and

Temple constantly runs "simulations," as she calls them, in her head: "I visualize the animal entering the chute, from different angles, different distances, zooming in or wide angle, even from a helicopter view—or I turn myself into an animal, and feel what it would feel entering the chute."

But if one thinks only in pictures, I could not help reflecting, one might not understand what nonvisual thinking was like, and one would miss the richness and ambiguity, the cultural presuppositions, the depth, of language. All autistics, Temple had said earlier, were intensely visual thinkers, like her. If this was true, was it, I wondered, more than a coincidence? Was Temple's intense visuality a vital clue to her autism?

A cattle farm, even a large one, is often a quiet place, but when we arrived we could hear a great tumult of bellowing. "They must have separated the calves from the cows this morning," Temple said, and, indeed, this was what had happened. We saw one cow outside the stockade, roaming, looking for her calf, and bellowing. "That's not a happy cow," Temple said. "That's one sad, unhappy, upset cow. She wants her baby. Bellowing for it, hunting for it. She'll forget for a while, then start again. It's like grieving, mourning—not much written about it. People don't like to allow them thoughts or feelings. Skinner wouldn't allow them."

As an undergraduate in New Hampshire, she had written to B. F. Skinner, the great behaviorist, and finally she had visited him. "It was like having an audience with God," she said. "It was a letdown. He was just a regular human being. He said, 'We don't have to know how the brain works—it's just a matter of conditioned reflexes.' No way *I* could believe it was just stimulus-response." The Skinner era, Temple concluded, was one that denied feelings to animals and rationalized regarding them as automata; it was an era of exceptional cruelty, both in

---

Somebody Somewhere) describes how she "adopted" two personas, Carol and Willie, and thought and spoke *through* them, in the many years when she had only a rudimentary identity herself.

animal experimentation and in the management of farms and slaughterhouses. She had read somewhere that behaviorism was an uncaring science, and this was exactly how she herself felt about it. Her own aspiration was to bring a vivid sense of animals' feelings back into husbandry.

Seeing the grieving cow and hearing the bereft bellows angered Temple and turned her mind toward inhumanities in slaughter. She had nothing to do with chickens, she said, but the killing of chickens was particularly loathsome. "When it's time for chickens to go to McNuggetland, they pick 'em up, hang 'em upside down, cut their throats." A similar shackling of cattle, and hanging them upside down so that the blood rushes to their heads before their throats are cut, is a common sight in old kosher slaughterhouses, she said. "Sometimes their legs get broken, they scream in pain and terror." Mercifully, such practices are now starting to change. Properly performed, "slaughter is more humane than nature," she went on. "Eight seconds after the throat's cut, endorphins are released; the animal dies without pain. It is similar in nature, after sheep have been ripped up by coyotes. Nature has done this to ease the pain of a dying animal." What is terrible, the more so because it is avoidable, she feels, is pain and cruelty, the introduction of fear and stress before the lethal cutting; and it is this that she is most concerned to prevent. "I want to reform the meat industry. The activists want to shut it down," she said, and added, "I don't like radical anything, left or right. I have a radical dislike of radicals."

Away from the bellowing of the separated calves and mothers, whose distress Temple seemed to feel in her bones, we found a calm, quiet area of the farm, where cattle were browsing placidly. Temple knelt and held out some hay, and a cow came over to her and took the hay, nudging her hand with its soft muzzle. A soft, happy look came over Temple's face. "Now I'm at home," she said. "When I'm with cattle, it's not at all cognitive. I know what the cow's feeling."

The cattle seemed to sense this, sensed her calm, her confidence, and came up to her hand. They did not come up to me, sensing, perhaps, the unease of the city dweller, who, living mostly in a world of cultural conventions and signals, is unsure how to behave with huge, nonverbal animals.

"It's different with people," she went on, repeating her earlier remark about feeling like an anthropologist on Mars. "Studying the people there, trying to figure out the natives. But I don't feel like that with animals."

I was struck by the enormous difference, the gulf, between Temple's immediate, intuitive recognition of animal moods and signs and her extraordinary difficulties understanding human beings, their codes and signals, the way they conduct themselves. One cannot say that she is devoid of feeling or has a fundamental lack of sympathy. On the contrary, her sense of animals' moods and feelings is so strong that these almost take possession of her, overwhelm her at times. She feels she can have sympathy for what is physical or physiological—for an animal's pain or terror—but lacks empathy for people's states of mind and perspectives.[7] When she was younger, she was hardly able to interpret even the simplest expressions of emotion; she learned to "decode" them later, without necessarily feeling them. (Similarly, Dr. Hermelin, in London, had told me a story about an intelligent autistic girl of twelve who came to her and said, of another student, "Joanie is making a funny noise." Upon going to investigate, Hermelin found Joanie crying bitterly. The meaning of weeping had been completely missed by the autistic girl: she had merely registered it as something physical, "a funny noise." I was reminded, too,

---

[7] She was deeply affected, physically shocked, when, during our talk, I imitated a young man with extremely severe Tourette's syndrome—how, with violent tics, he had put out his own eyes. Expressions of raw impulse, violence, pain, she perceived, reacted to, straightaway. I was reminded of how, in a completely benign way, Shane, with his Tourette's, had got through to the autistic children at Camp Winston, at a level of emotion and animal sympathy, a level more elemental, more directly conveyable, than that of complex states of mind and perspectives.

of Jessy Park, and how she was fascinated by the fact that onions could make one weep but was totally unable to comprehend that one could also weep for joy.)[8]

"I can tell if a human being is angry," she told me, "or if he's smiling." At the level of the sensorimotor, the concrete, the unmediated, the animal, Temple has no difficulty. But what about children, I asked her. Were they not intermediate between animals and adults? On the contrary, Temple said, she had great difficulties with children—trying to talk with them, to join in their games (she could not even play peekaboo with a baby, she said, because she would get the timing all wrong)—as she had had such difficulties herself as a child. Children, she feels, are already far advanced, by the age of three or four, along a path that she, as an autistic person, has never advanced far on. Little children, she feels, already "understand" other human beings in a way she can never hope to.

What is it, then, I pressed her further, that goes on between normal people, from which she feels herself excluded? It has to do, she has inferred, with an implicit knowledge of social conventions and codes, of cultural presuppositions of every sort. This implicit knowledge, which every normal person accumulates and generates throughout life on the basis of experience and encounters with others, Temple seems to be largely devoid of. Lacking it, she has instead to "compute" others' intentions and states of mind, to try to make algorithmic, explicit, what for the rest of us is second nature. She herself, she infers, may never have had the normal social experiences from which a normal social knowledge is constructed.

And it may be from this, too, that her difficulties with ges-

---

[8] Some autistic people keep dogs, as blind or deaf people may do, to assist their perceptions—in this case, social perceptions. They may use dogs to "read" the minds and intentions of visitors, which they may feel unable to do themselves. I know two autistic people who regard their dogs as having "telepathic" abilities, but of course the abilities of their dogs are merely normal canine ones—and indeed normal human ones—which they themselves lack.

ture and language stem—difficulties that were devastating when she was a near-speechless child, and also in the early days of speech, when she mixed all her pronouns up, not able to grasp the different meanings of "you" and "I," depending on context.

It is extraordinary to hear Temple speak of this time, or to read of it in her book. When she was three, as an outside chance, although her family did not have much belief in its promise, she was sent to a special nursery school for disturbed and handicapped children, and a trial of speech therapy was suggested. Somehow, the school and the speech therapist got through to Temple, rescued her (she later came to feel) from the abyss, and started her on her slow emergence. She remained clearly autistic, but her new powers of language and communication now gave her an anchor, some ability to master what had been total chaos before. Her sensory system, with its violent oscillations of oversensitivity and undersensitivity, started to stabilize a little. There were many periods of backsliding and regression, but it is clear that by the age of six she had achieved fair language and, with this, had crossed the Rubicon that divides high-functioning people like her from low-functioning ones, who never achieve proper language or autonomy. With the access of language, the terrible triad of impairments—social, communicative, and imaginative—began to yield somewhat. Temple started having some contact with others, especially one or two teachers who could appreciate her intelligence, her specialness, and could withstand her pathology—her now-incessant talking and questioning, her strange fixations, her rages. No less crucial was the emergence of some genuine playfulness and creativity—painting, drawing, making cardboard models and sculptures, as well as "unique and creative ways of being naughty." At eight, Temple was starting to achieve the pretend-play that normal children achieve as toddlers, but the lower-functioning autistic child never achieves at all.

Her mother, an aunt, and several teachers were crucial, but also crucial, on the long journey up, was the slow development that many autistics show; autism, being a developmental disorder, tends to become less extreme as one grows older, and one may learn to cope with it better.

Temple had longed for friends at school and would have been totally, fiercely loyal to a friend (for two or three years, she had an imaginary friend), but there was something about the way she talked, the way she acted, that seemed to alienate others, so that, while they admired her intelligence, they never accepted her as part of their community. "I couldn't figure out what I was doing wrong. I had an odd lack of awareness that I was different. I thought the other kids were different. I could never figure out why I didn't fit in." Something was going on between the other kids, something swift, subtle, constantly changing—an exchange of meanings, a negotiation, a swiftness of understanding so remarkable that sometimes she wondered if they were all telepathic. She is now aware of the existence of these social signals. She can infer them, she says, but she herself cannot perceive them, cannot participate in this magical communication directly, or conceive the many-leveled kaleidoscopic states of mind behind it. Knowing this intellectually, she does her best to compensate, bringing immense intellectual effort and computational power to bear on matters that others understand with unthinking ease. This is why she often feels excluded, an alien.

A crucial event occurred when she was fifteen. She had become fascinated with the squeeze chutes used to hold cattle. A science teacher took her fixation seriously, instead of scoffing, and suggested she actually build her own squeeze chute. From this beginning, he guided her from particular considerations of farm animals and machinery to a general interest in biology and all science. And here Temple, still quite abnormal in her understanding of ordinary or social language—she still missed allusions, presuppositions, irony, metaphors, jokes—found the language of science and technology a huge relief. It was much clearer, much more explicit, with far less depending on unstated assumptions. Technical language was as easy for her as social language was difficult, and it now provided her with an entry into science.

But if there was a resolution at this level, with the focusing of much of her intellectual and emotional energy on science, other tensions, anxieties—even agonies—remained. With the onset of adolescence, Temple started to confront the realization that she might never lead a "normal" life, or enjoy the "normal" satisfactions—love and friendship, recreation and society—that went with it. This realization may be devastating for gifted young autistic people at this stage and has been a cause of depression in some and even of suicide on occasion. Temple dealt with this realization partly by renunciation and dedication: she would be celibate, she decided, and would make science her whole life.

Adolescence also taught her that not only her emotional state but her whole mental and physical being were very finely tuned and could easily be thrown out of balance by certain sensory stimuli, stress, exhaustion, or conflict.[9] The hormonal turbulences of adolescence, in particular, threw her up and down. But there was also a passion, an intensity, at this turbulent time; and it was only when she had finished college and was launched on her career, she said, that she could afford to calm down. Indeed, she felt she had to; otherwise her body would destroy itself. At this point, she started on a small dose of imipramine, a drug marketed as an antidepressant. In her book, Temple speaks of the pros and cons of this:

Gone are the frenzied searches for the basic meaning of life. I no longer fixate on one thing since I am no longer driven. During the last four years I have written very few entries in my diary because the anti-depressant has taken away much of the fervor. With the passion subdued, my career and . . . business is going well. Since I am more re-

---

[9] The provocative stimuli may be very different from one person to another: one autistic person will be intolerant of high-pitched noises, another of low-pitched noises, one of a fan, another of a washing machine. There may also be various visual, tactile, and olfactory idiosyncrasies.

laxed, I get along better with people, and stress-related health problems, such as colitis, are gone. Yet if medication had been prescribed for me in my early twenties, I might not have accomplished as much as I have. The "nerves" and the fixations were great motivators until they tore my body apart with stress-related health problems.

I was reminded, reading this, of what Robert Lowell once told me about being on lithium for his manic-depressive disorder: "I feel much 'better,' in a way, calmer, stabler—but my poetry has lost much of its force." While Temple, too, is well aware of the cost of being calmed down, she feels, at this point in her life, that it is well worth paying. Yet she sometimes misses the emotions, the frenzies, she once felt.

The other side of a much-retarded development may be a continuing ability to develop social skills and perceptions throughout life, and the last twenty years have indeed been years of continuing development for Temple. Ten years ago, when she first started lecturing, I had been told, she often seemed not to be addressing the audience—she would have no eye contact and might actually be facing in another direction—and she could not take questions after the lecture. Now she spends almost 90 percent of her time on the road, lecturing around the world, sometimes about autism, sometimes about animal behavior. She has become much more fluent in her lecturing style, has more eye contact with the audience, and may even add humorous asides and improvisations; she answers—and, if need be, parries—questions easily. In her social life, she seems also to have developed, so that most recently, Temple told me, she has been able to enjoy spending time with two or three friends. But achieving genuine friendship, appreciating other people for their otherness, for their own minds, may be the most difficult of all achievements for an autistic person. Uta Frith, in *Autism and Asperger Syndrome*, writes, "Asperger syndrome individuals . . . do not

seem to possess the knack of entering and maintaining intimate two-way personal relationships, whereas routine social
interactions are well within their grasp." Her colleague Peter
Hobson writes of an intelligent but autistic man who could
not comprehend the meaning of "a friend." Yet it seemed to
me, as I listened to her, that Temple, now in her forties, had
grasped at least something of the nature of friendship.

On this note—we had been walking and talking for almost two hours—we finished our visit to the university farm
and took a break for lunch. Temple, it seemed to me, was
happy to stop talking, stop thinking for a while; there had
been an almost ferocious intensity in the self-examination I
had forced on her (although it was not unlike the self-
examination she forces on herself daily, struggling, as always,
to understand and live with autism in a nonautistic world).
"Normality" had been revealed more and more, as we spoke,
as a sort of front, or facade, for her, albeit a brave and often
brilliant front, behind which she remained, in some ways, as
far "outside," as unconnected, as ever. "I can really relate to
Data," she said as we drove away from the farm. She is a "Star
Trek" fan, as I am, and her favorite character is Data, an an-
droid who, for all his emotionlessness, has a great curiosity,
a wistfulness, about being human. He observes human behavior minutely, and sometimes impersonates it, but longs,
above all, to *be* human. A surprising number of people with
autism identify with Data, or with his predecessor, Mr.
Spock.

This was the case with the B.'s, the autistic family I had visited in California—the older son, like the parents, with
Asperger's syndrome, the younger with classical autism.
When I first arrived at their house, the whole atmosphere was
so "normal" that I wondered if I had been misinformed, or if
I had not, perhaps, ended up at the wrong house, for there was
nothing obviously "autistic" about them or it. It was only after I had settled down that I noticed the well-used trampoline,

where the whole family, at times, likes to jump and flap their arms; the huge library of science fiction;[10] the strange cartoons pinned to the bathroom wall; and the ludicrously explicit directions, pinned up in the kitchen—for cooking, laying the table, and washing up—suggesting that these had to be performed in a fixed, formulaic way (this, I learned later, was an autistic in-joke). Mrs. B. spoke of herself, at one point, as "bordering on normality," but then made clear what such "bordering" meant: "We know the rules and conventions of the 'normal,' but there is no actual transit. You act normal, you learn the rules, and obey them, but . . ."

"You learn to ape human behavior," her husband interpolated. "I still don't understand what's behind the social conventions. You observe the front—but . . ."

The B.'s, then, had learned a front of normality, which was necessary, given their professional lives, their living in the suburbs and driving a car, their having a son in regular school, and so on. But they had no illusions about themselves. They recognized their own autism, and they had recognized each other's, at college, with a sense of such affinity and delight that it was inevitable they would marry. "It was as if we had known each other for a million years," Mrs. B. said. While they were well aware of many of the problems of their autism, they had a respect for their differentness, even a pride. Indeed, in some autistic people this sense of radical and ineradicable

---

[10] Many high-functioning autistic people describe a great fondness for, almost an addiction to, alternative worlds, imaginary worlds such as those of C. S. Lewis and Tolkien, or worlds they imagine themselves. Thus both the B.'s and their older son have spent years constructing an imaginary world with its own landscapes and geography (endlessly mapped and drawn), its own languages, currencies, laws, and customs—a world in which fantasy and rigidity play equal parts. Thus days might be spent computing the total grain production or silver reserves in Leutheria, or designing a new flag, or calculating the complex factors determining the value of a thog—this occupies hours of the B.'s leisure time at home together, Mrs. B. providing the science and technology; Mr. B. the politics, languages, and social customs; and their son the natural features of the often-warring countries.

differentness is so profound as to lead them to regard themselves, half jokingly, almost as members of another species
("They beamed us down on the transporter together," as the
B.'s liked to say), and to feel that autism, while it may be seen
as a medical condition, and pathologized as a syndrome, must
also be seen as a whole mode of being, a deeply different mode
or identity, one that needs to be conscious (and proud) of
itself.

Temple's attitudes seem similar to this: she is very aware (if
only intellectually, inferentially) of what she is missing in life,
but equally (and directly) aware of her strengths, too—her concentration, her intensity of thought, her single-mindedness,
her tenacity; her incapacity for dissembling, her directness,
her honesty. She suspects—and I, too, was coming more and
more to suspect—that these strengths, the positive aspects of
her autism, go with the negative ones. And yet there are times
when she needs to forget that she is autistic, to feel at one
with others, not outside, not different.

Having spent the morning among beef cattle, and plan-
ning to visit a slaughterhouse (or "meat-packing plant," in the
industry's euphemism) in the afternoon, we found ourselves a
little averse to meat and had a Mexican meal of rice and
beans. After lunch, we drove to the airport and took a tiny
commuter plane, then drove out to the plant. Temple was
proud of its layout and wanted to show me how it looked.
Such plants are closed to the public and maintain a high degree of security. Temple had designed the facilities a couple of
years earlier and still had her overalls and I.D. with the plant's
insignia. But I was a problem: What was to be done with me?
Temple had thought of this in the morning and had selected
from her hat collection a sanitary engineer's bright-yellow
hard hat. She handed it to me, saying, "That'll do. You look
good in it. It goes with your khaki pants and shirt. You look
exactly like a sanitary engineer." (I blushed; no one had ever
told me this before.) "Now all you have to do is behave like
one, think like one." I was astounded at this, for autistic peo-

ple, it is said, have no pretend-play, and here Temple had, very coolly, and without the slightest hesitation, determined on a subterfuge and was all set to smuggle me into the plant.

Our entry, in the event, went off without trouble. Temple drove through the gate with a sublime air of confidence, waved cheerily to the security guard, and was as cheerily waved in. "Keep the hard hat on," she said to me when we parked. "Keep it on the whole time. You're a sanitary engineer here."

We stopped to lean over the fence where the cattle are corralled outside the large plant building and then followed the path that the cattle follow when they go on their last journey, up and up a curving ramp leading into the main plant building—"the stairway to Heaven," Temple called it. Here, again, I was puzzled. The autistic have difficulty with metaphor, it is said, and never use irony. But, looking at Temple's straight, serious expression, I was not sure that, for her, this was metaphor or irony. She had heard the phrase—perhaps it seemed to her literally true. She describes in her autobiography a similar literalization of a symbol when, as an adolescent, she heard a minister quote John 10:9—"I am the door: by me if any man enter in, he shall be saved"—and the minister added, "Before each of you there is a door opening into Heaven. Open it and be saved." Temple writes:

> Like many autistic children, everything was literal to me. My mind centered on one thing. Door. A door opening to Heaven. . . . I had to find that door. . . . The closet door, the bathroom door, the front door, the stable door—all were scrutinized and rejected as the door. Then one day . . . I noticed that an addition to our dorm was being constructed. . . . A small platform extended out from the building and I climbed on it. And there was the door! It was a little wooden door that opened out onto the roof. . . . A feeling of relief flooded me. . . . A feeling of love and joy . . . I'd found it! The door to my Heaven.

Later, Temple told me that she believed in some sort of existence after death (even if it was only as "an energy impression" in the universe). Intensely conscious of animals'
emotions, their "humanity," she had to grant them some sort
of immortality, too.

We walked slowly up by the side of the gently curving,
high-walled ramp, where cattle walk in single file, blithely unconscious of what is to come, up to the stunner, with its lethal bolt. Temple has been a pioneer in the design of such
ramps, and her name is associated, in the trade, with the introduction of curved chutes. As we ascended the catwalk,
looking over the chute's walls, Temple told me of their special
virtues, how curved chutes prevented the animals from seeing
what was at the other end of the ramp until they were almost
there (thus preventing any apprehension) and, at the same
time, took advantage of the cow's natural tendency to circle.
The high walls prevented upsetting distractions and served to
concentrate the animals on their walk.

At the top of the ramp, inside the building, the animals
found themselves moved, almost insensibly, onto a conveyor
belt running under their bellies. (This "double-rail restrainer"
was another innovation of Temple's.) A few seconds later, the
animal is instantly killed by a bolt shot by compressed air
through the brain. A very similar system, Temple told me,
might be used for hogs as well, though typically these would
be killed by electrical stunning, not a bolt. She added an interesting gloss: "An electroshock machine"—such as is used in
some psychiatric facilities—"and a hog stunner have almost
exactly the same parameters: around one ampere, at three
hundred volts." A slight misplacement of the leads, she added,
and the patient would be killed, stunned, like a hog. She was
a bit shocked, she allowed, when she realized this.

I got a sense of horror as Temple showed me the stunner,
but the cattle, she assured me, had no intimation, no apprehension, of what was to happen to them; her whole effort, indeed, was to remove anything that could frighten or stress the

animals, so that they could go peacefully, gently, unknowingly, to their death. But I still felt queasy about the whole thing. How did she feel, how did others feel, working in such places?

Temple has explored this and has written a classic paper on the subject.[11] Some employees in slaughterhouses, she notes, rapidly develop a protective hardness and start killing animals in a purely mechanical way: "The person doing the killing approaches his job as if he was stapling boxes moving along a conveyor belt. He has no emotions about his act." Others, she reveals, "start to enjoy killing and . . . torment the animals on purpose." Speaking of these attitudes turned Temple's mind to a parallel: "I find a very high correlation," she said, "between the way animals are treated and the handicapped. . . . Georgia is a snake pit—they treat [handicapped people] worse than animals. . . . Capital-punishment states are the worst animal states and the worst for the handicapped."

All this makes Temple passionately angry, and passionately concerned for humane reform: she wants to reform the treatment of the handicapped, especially the autistic, as she wants to reform the treatment of cattle in the meat industry. (The only fitting approach to killing animals, the only one that shows respect for the animal, Temple feels, is the ritual or "sacred" one.)

It was an enormous relief getting out of the slaughter plant, away from the hideous smell, which seemed to permeate every inch of the place and had made me hold my stomach and my breath sometimes in an effort not to puke; an enormous relief, once we were outside, to breathe the sharp, clear air, untainted with the smell of blood and offal; an enormous relief, morally, to get away from the idea of killing. I asked Temple about this as we drove away. "Nobody should kill an-

---

[11] Her article, "Behavior of Slaughter Plant and Auction Employees Toward the Animals," appeared in *Anthrozoos: A Multidisciplinary Journal on the Interactions of People, Animals, and Environment* in the spring of 1988.

imals all the time," she said, and she told me she had written much on the importance of rotating personnel, so that they would not be constantly employed in killing, bleeding, or driving. She herself is in need of other atmospheres and occupations, and these form a vital and altogether pleasanter part of her life. Her understanding of the psychology and behavior of herd animals is sought not only by feedlots and slaughterhouses all over the world but by sheep shearers as far away as New Zealand, and by game parks and zoos. I had the feeling that she might like to spend time on the African veldt, as a consultant on elephant herds and prey animals like antelopes and wildebeest. But would she, I wondered, be able to understand apes (who have some "theory of mind") as well as she understood cattle? Or would she find them bewildering, impenetrable, the way she found children and other human beings? ("With farm animals, I feel their behavior," she said later. "With primates I intellectually understand their interactions.")

Temple's deepest feelings are for cattle; she feels a tenderness, a compassion, for them that is akin to love. She spoke of this at length as we made our way to our next destination, a feedlot—how she sought gentleness, holding cattle in the chute, how she sought to transmit calmness to the animals, to bring them peace in the last moments of their lives. This, for her, is half-physical, half-sacred, this cradling of an animal in the last moments of its life, and it is something she endlessly tries to teach the people who operate the chutes in the slaughter plants. She told me a story of how one plant manager, while very defensive about being advised on this by her, was fascinated by her power to calm excited animals, and how, unknown to her, he had spied on her through a hole in the ceiling as she worked. This had occurred when she was consulting at a slaughterhouse in the South, and the entire scene, and its context, kept returning to her mind: she told me the story half a dozen times in the afternoon, each time at length, and in virtually the same words.

I was struck both by the vividness of the reexperience, the

memory, for her—it seemed to play itself in her mind with extraordinary detail—and by its unwavering quality.[12] It was as if the original scene, its perception (with all its attendant feelings), was reproduced, replayed, with virtually no modification. This quality of memory (so akin to Stephen Wiltshire's, in a way) seemed to me both prodigious and pathological—prodigious in its detail and pathological in its fixity, more akin to a computer record than to anything else. Such computational analogies, indeed, are frequently brought up by Temple herself: "My mind is like a CD-ROM in a computer—like a quick-access videotape. But once I get there, I have to play that whole part." She could not just focus, for instance, on the cradling of an animal in its last moments; she had to play, in memory, the entire scene, from the animal entering the chute and progressing steadily ("no fast-forward, it takes about two minutes") until the death of the animal and its collapse, after

---

[12] The psychologist Frederic Bartlett writes of remembering as "reconstruction," but for Temple (as for Stephen), seemingly, this does not occur, or occurs to a much smaller extent than usual. Nor is memory, for her, entirely internalized as part of the self—thus her frequent allusions to "videotapes" and "computer records," and other external forms of memory storage.

Temple's self-description here is intriguingly at odds with some of the current formulations of imagery and memory, as conceived by Damasio, Edelman, and others. Thus Damasio writes, in *Descartes' Error*:

Images are *not* stored as facsimile pictures of things, or events, or words, or sentences. The brain does not file Polaroid pictures of people, objects, landscapes; nor does it store audiotapes of music and speech; it does not store films of scenes in our lives. . . . In brief, there seems to be no permanently held pictures of anything, even miniaturized, no microfiches or microfilms, no hard copies.

Yet this, Damasio emphasizes, "must be reconciled with the sensation . . . that we *can* conjure up" such reproductions or facsimile images. One must wonder, if this is the case, whether Temple—and also Franco and Stephen (and Luria's Mnemonist)—are merely, like the rest of us, susceptible to an *illusion* of reproduction, or whether in fact (as Jerome Bruner suggests) there may be in them some failure of integration of perceptual systems with higher integrative ones, and with concepts of self, so that *relatively* unprocessed, uninterpreted, unrevised images persist.

its throat has been cut. "I can do anything the computers in *Jurassic Park* do," she continued. "I can do all that stuff in my head. . . . I actually have that machine in my head. I run it in my mind. I play the tape—it's a slow method of thinking." But an ideal sort of thinking for much of her work. She designs the most elaborate facilities in her mind, visualizing every component of the system, juxtaposing them in different ways, viewing them from different angles, from near and far. Once the design is complete, she will "run a simulation" in her mind— that is, imagine the entire plant in operation. This simulation may show an unexpected problem, and when this happens she will pinpoint the problem, modify the design, do another simulation—several simulations, if need be—until the design is perfect. Only now, when all is clear in her mind, does she make an actual blueprint of it. No more attention is needed at this point; the rest is mechanical. "Once I get the basic thing laid out, I just put it on paper. I can listen to the TV. There's no emotion in it. I just turn on my Sun workstation and do it."

But this sort of simulation or concrete imagery is much less appropriate when she has to do other kinds of thinking— symbolic or conceptual or abstract thinking. To understand the proverb "A rolling stone gathers no moss," she said, "I have to run a video of the rock rolling and getting the moss off before I can think of what it 'means.' " She has to concretize before she can generalize. At school, she could not understand the Lord's Prayer until she "saw" it in concrete images: " 'The power and the glory' were high-tension electric wires and a blazing sun; the word 'trespass' . . . a 'No Trespassing' sign on a tree."[13]

In her autobiography, and, more concisely, in a thirty-page

---

[13] When Temple lectures, she often uses very odd slides, mixed in with the usual diagrams and charts—slides that might bear no discernible relation to her theme and might convey nothing to her audience, since in fact they are designed not for them but for her, private jottings or mnemonics for her own trains of thought. For instance, a joke slide of a roll of toilet paper made from sandpaper reminds her to speak about tactile sensitivity in autism.

article published a little before the book—"My Experiences as an Autistic Child," which appeared in the *Journal of Ortho-molecular Psychiatry* in 1984—Temple indicates how, even as a child, she scored at the top of the recorded norms in spatial tests and visual tests but did rather badly in abstract and se-quential tasks. (Such "profiles" are characteristic of autistic people: they tend to show "scatter," or extreme unevenness, on so-called intelligence tests.) In some cases, Temple writes, the scores were misleading, because tasks that might have been very difficult for her if she had done them in the "nor-mal" way were easy because she did them in an idiosyncratic, visual way: thus sentences and poems, and strings of numbers, instantly generated visual images, and these were what she re-membered, not the words or numbers as such. Complex calcu-lations, impossible for her in the normal way, might become possible if she transformed them into visual images.[14]

Visual thinking in itself is not abnormal, and Temple was quick to point out that she knows several nonautistic people—engineers, designers—who seem able to "see" what they need to do, to make designs in their mind and test them in simulations, just as she does.[15] Indeed, she often gets on very well with such people, especially her friend Tom. He is a

---

[14] As Temple described this and gave examples, I was reminded of the Mnemon-ist described by A. R. Luria (in *The Mind of a Mnemonist*) and his bizarre, purely visual way of transforming words and numbers into images. The Mnemonist, in-deed, thought exclusively in images—and sometimes overwhelmingly; hundreds of these might be generated in the course of listening to a single paragraph or a short poem. Thinking in images gave him great strength—provided, in Luria's words, "a powerful base on which to operate, allowing him to carry out in his mind manipulations which others could only perform with objects." But such thinking also created strange difficulties, sometimes preposterous ones, when it could not be replaced by verbal-logical thought. Luria's Mnemonist was not in the least autistic, but his visual thought processes—his concrete imagery, at least—were remarkably close to Temple's and perhaps shared a similar physiological ba-sis. She was fascinated when I told her of the Mnemonist and felt that her thinking was indeed very similar to his.

[15] Precisely such a mode of mind was possessed by the great inventor Nikola Tesla: "When I get an idea I start at once building it up in my imagination. I

powerful, creative visualizer, like her, and is also, like her, unorthodox, roguish, fond of pranks. "I get on the same wavelength as Tom," Temple said, "though it's a childish wavelength." But, above all, she enjoys working with Tom—this, too, is "childish," but a form of childishness that is essentially creative. "Tom and I are little children," she said. "Concrete is grown-up mud, steel is grown-up cardboard, building is grown-up play."

I was moved by Temple's words, with their lovely analogizing of creativity and child's play, and thought what a healthy development this had been in her. And moved, too, when she spoke of her relation to Tom. I wondered whether indeed she loved him and had ever thought of a sexual relationship or marriage with him. I asked her about this—asked whether she had ever had sexual relationships, or dated, or fallen in love.

No, she said. She was celibate. Nor had she ever dated. She found such interactions completely baffling and too complex to deal with; she was never sure what was being said, or implied, or asked, or expected. She did not know, at such times, where people were coming from, or their assumptions or presuppositions, or intentions. This was common with autistic people, she said, and one reason why, though they had sexual feelings, they rarely succeeded in dating or having sexual relationships.

But the problem was not just in actual dating or relating. "I have never fallen in love," she told me. "I don't know what it's like to rapturously fall in love."

"What do you imagine 'falling in love' is like?" I asked.

"Maybe it's like swooning—if not that, I don't know."

I thought the phrase "falling in love," with its suggestion of overwhelming feeling or transports, might be the wrong term to use. I amended my question to "What is 'loving'?"

---

change the construction, make improvements and operate the device in my mind. It is absolutely immaterial to me whether I run my turbine in my thought or test it in my shop. *I even note if it is out of balance.*"

"Caring for somebody else . . . I think gentleness would have something to do with it."

"Have you cared for somebody else?" I asked her.

She hesitated for a moment before answering. "I think lots of times there are things that are missing from my life."

"Is this painful?"

"Yeah . . . I guess." Then she added, "When I started holding the cattle, I thought, What's happening to me? Wondered if that was what love is . . . it wasn't intellectual anymore."

She is wistful about love, in a sense, but cannot actually imagine how it might be to feel passion for another person. "I couldn't understand how my roommate would swoon over our science teacher," she recalled. "She was overwhelmed with emotion. I thought, He's nice, I can see why she likes him. But there was no more than that."

The capacity to "swoon," to experience a passionate emotional response, seems diminished in other areas, too—not merely in relation to other people. For, after speaking of her roommate, Temple immediately said, "It's similar with music—I don't swoon." She has absolute pitch, she added (this is normally very rare, but is relatively common in people with autism), and a precise and tenacious musical memory, but, on the whole, music fails to move her. She finds it "pretty," but it evokes nothing deep in her, only literal associations: "Whenever I hear that *Fantasia* music, I see those stupid dancing hippos." It doesn't seem to "call" her. She doesn't "get" music, she said—doesn't see what it is "about." One might suppose that Temple is simply not "musical," despite her absolute pitch and her ear. But her inability to respond deeply, emotionally, subjectively, is not confined to music. There is a similar poverty of emotional or aesthetic response to most visual scenes: she can describe them with great accuracy but they do not seem to correspond to or evoke any strongly felt states of mind.

Temple's own explanation of this is a simple mechanical one: "The emotion circuit's not hooked up—that's what's wrong." For the same reason, she does not have an unconscious, she says; she does not repress memories and thoughts, like normal people. "There are no files in my memory that are repressed," she asserted. "You have files that are blocked. I have none so painful that they're blocked. There are no secrets, no locked doors—nothing is hidden. I can infer that there are hidden areas in other people, so that they can't bear to talk of certain things. The amygdala locks the files of the hippocampus. In me, the amygdala doesn't generate enough emotion to lock the files of the hippocampus."

I was taken aback and said, "Either you are incorrect or there is an almost unimaginable difference of psychic structure. Repression is universal in human beings." But, having said it, I was not so sure. I could imagine organic conditions in which repression might fail to develop, or be destroyed, or be overwhelmed. This seems to have been the case with Luria's Mnemonist, who, though not autistic, had memories of such vividness as to be inextinguishable—even though some of these were so painful that they would surely have been repressed had this been (physiologically) possible. I myself had had a patient in whom damage to the frontal lobes of the brain "released" some of the most deeply repressed memories— memories of a murder he had committed—and forced them upon his terror-stricken consciousness.

I had another patient, an engineer, with massive frontal lobe damage from a hemorrhage, whom I would often see reading *Scientific American*. He was still well able to understand most of the articles, but he said that they no longer evoked any sense of wonder in him—the very sense that, formerly, had been central to his passion for science.

Another man, a former judge who is described in the neurological literature, had frontal lobe damage from shell fragments in the brain, and, in consequence, found himself totally deprived of emotion. It might be thought that the absence of emotion, and of the biases that go with it, would have rendered him more impartial—indeed, uniquely qualified—as a judge. But he himself, with great insight, resigned from the bench, saying that he could no longer enter sympathetically into the motives of anyone concerned, and that since justice involved feeling, and not merely thinking, he felt that his injury totally disqualified him.[16]

Such cases show us how the whole affective basis of life can be undercut by neurological damage. But there is something much more selective about the affective problems in autism; there is by no means an overall flatness or blandness, despite Temple's comments about the "emotion circuit" or amygdala. An autistic person can have violent passions, intensely charged fixations and fascinations, or, like Temple, an almost overwhelming tenderness and concern in certain areas. In autism, it is not affect in general that is faulty but affect in relation to complex human experiences, social ones predominantly, but perhaps allied ones—aesthetic, poetic, symbolic, etc. No one, indeed, brings this out more clearly than Temple herself.

Both as a person struggling to understand herself and as a scientist exploring animal behavior, Temple is constantly exercised by her own autism, constantly seeks models or similes to understand it. She feels that there is something mechanical about her mind, and she often compares it to a computer, with many elements in parallel (a parallel-distributed processor, to use the technical term), seeing her own thinking as "computation" and her memory as computer files. She surmises that her mind is lacking some of the "subjectivity," the inwardness, that others seem to have. She sees the elements of her thoughts as concrete and visual images, to be permuted or associated in different ways.[17] She believes that the visual parts

---

[16] The founding of reason on feeling is the central theme of Antonio Damasio's book, *Descartes' Error*.

[17] Temple's self-description here made me think of Coleridge's delineation of Fancy: "[It] has no other counters to play with, but fixities and definites. . . . [It] must receive all its materials ready made from the law of association." I think that the overwhelming tendency to fixed, concrete, perceptual images, and their quasi-mechanical association, permutation and play—which one sees in autism and sometimes Tourette's syndrome—while it may dispose to vivid and active

of her brain and those concerned with processing a great mass of data simultaneously are very highly developed, and that this is generally so in autistic people, and she believes that the verbal parts of her brain, and those designed for sequential processing, are comparatively underdeveloped, and that this, too, is very common in autistic people.[18] She is conscious of the "stickiness" of attention in herself, so that there is great tenacity on the one hand but a lack of agility and pliability on the other; she ascribes this to a defect in her cerebellum, the fact that (as an MRI has shown) it is below normal size in her. She believes such cerebellar defects are significant in autism, though scientific opinion is divided on this.

She feels that there are usually genetic determinants in autism; she suspects that her own father, who was remote, pedantic, and socially inept, had Asperger's—or, at least, autistic traits—and that such traits occur with significant frequency in the parents and grandparents of autistic children.[19] Though she feels early environment (in pigs or people) plays a crucial role in psychic development, she does not hold (as Bruno Bettelheim did) that parental behavior is responsible for autism; it is more likely, she thinks, that autism itself presents barriers to contact and communication that parents may be unable to penetrate, so that the entire range of sensory and social experiences (especially holding and deep pressure) becomes severely impoverished.

Fancy (in Coleridge's sense), may also dispose against Imagination (as he calls it, in contrast), which "dissolves, diffuses, dissipates, in order to recreate." The creation, or re-creation, of the Imagination entails a letting-go of fixities and definites in order to revise and reconstruct—and it is just this that seems so difficult in the overprecise and rigid mind of an autistic person.

[18] Russell Hurlburt, at the University of Nevada, has studied the ways in which individuals report or represent their inner experiences, their streams of thought. He has found that whereas normal (and neurotic or schizophrenic) subjects seem to utilize a combination of different modes—inner speech and hearing, feelings, bodily sensations, as well as visual images—subjects with Asperger's syndrome seem to use visual images exclusively or predominantly.

[19] That this is indeed the case has recently been shown by Ed and Riva Ritvo of UCLA.

Temple's own formulations and explanations generally correspond with the range of existing scientific ones, except that her emphasis on the necessity of early hugging and deep pressure is very much her own—and, of course, has been a mainspring in directing her thoughts and actions from the age of five. But she thinks that there has been too much emphasis on the negative aspects of autism and insufficient attention, or respect, paid to the positive ones. She believes that, if some parts of the brain are faulty or defective, others are very highly developed—spectacularly so in those who have savant syndromes, but to some degree, in different ways, in all individuals with autism. She thinks that she and other autistic people, though they unquestionably have great problems in some areas, may have extraordinary, and socially valuable, powers in others—provided that they are allowed to be themselves, autistic.

Moved by her own perception of what she possesses so abundantly and lacks so conspicuously, Temple inclines to a modular view of the brain, the sense that it has a multiplicity of separate, autonomous computational powers or "intelligences"—much as the psychologist Howard Gardner proposes in his book *Frames of Mind*. He feels that while the visual and musical and logical intelligences, for instance, may be highly developed in autism, the "personal intelligences," as he calls them—the ability to perceive one's own and others' states of mind—lag grossly behind.

Temple is impelled by two drives: a theorizing part of herself, which makes her want to find some general explanation of autism, some key that will be applicable to all of its phenomena and to every case; and a practical, empirical part of herself, which constantly faces the range and irreducible complexity and unpredictability of her own disorder, and the great range of phenomena in other autistic people, too. She is fascinated by the cognitive and existential aspects of autism and their possible biological basis, even though she is intensely aware that they are only part of the syndrome. She herself faces, almost every day, extreme variations, from overresponse to nonresponse, in her own sensory system, which cannot be explained, she feels, in terms of "theory of mind." She herself was already asocial at the age of six months and stiffened in her mother's arms at this time, and such reactions, common in autism, she also finds inexplicable in terms of theory of mind. (No one supposes that even normal children develop a theory of mind much before the age of three or four.) And yet, given these reservations, she is strongly attracted by Frith and other cognitive theorists; by Hobson and others who see autism as foremost a disorder of affect, of empathy; and by Gardner and his theory of multiple intelligences. Perhaps, indeed, all these theories, despite their different emphases, hover about the same point.

Temple has dipped into the chemical and physiological and brain-imaging researches on autism and emerged with the sense that they are still, at this point, fragmentary and inconclusive. But she holds to her notion of impaired "emotion circuits" in the brain, and she imagines these serve to link the phylogenetically ancient, emotional parts of the brain—the amygdala and the limbic system—with the most recently evolved, specifically human parts of the prefrontal cortex. Such circuits, she accepts, may be necessary to allow a new, "higher" form of consciousness, an explicit concept of one's self, one's own mind, and of other people's—precisely what is deficient in autism.

At a recent lecture, Temple ended by saying, "If I could snap my fingers and be nonautistic, I would not—because then I wouldn't be me. Autism is part of who I am." And because she believes that autism may also be associated with something of value, she is alarmed at thoughts of "eradicating" it. In a 1990 article she wrote:

> Aware adults with autism and their parents are often angry about autism. They may ask why nature or God created such horrible conditions as autism, manic depres-

sion, and schizophrenia. However, if the genes that caused these conditions were eliminated there might be a terrible price to pay. It is possible that persons with bits of these traits are more creative, or possibly even geniuses. . . . If science eliminated these genes, maybe the whole world would be taken over by accountants.

Temple arrived to pick me up at the hotel at exactly eight o'clock on Sunday morning, bringing along some additional articles of hers. I had the feeling that she was incessantly at work, that she used every available moment, "wasted" very little time, that virtually her entire waking life consisted of work. She seemed to have no recreations, no leisure. Even the weekend she had "scheduled" for me was by no means regarded as a social one but as forty-eight hours allocated for a special purpose, forty-eight hours set aside to allow a brief, intensive investigation of an autistic life, her own. If she sometimes saw herself as an anthropologist on Mars, she could see me as a sort of anthropologist, too, an anthropologist of autism, of her. She saw that I needed to observe her in all possible contexts and situations, amass a sufficient database to make correlations, to arrive at some general conclusions. That I might see with a sympathetic or friendly eye as well as an anthropological, one did not at first occur to her. So our visit was seen as work, and work to be carried through with the same conscientiousness and scrupulousness as all her work. Though in the normal course of events she invites people to her house, she would ordinarily never have shown her bedroom to a visitor; much less displayed, and illustrated the use of, the squeeze machine by her bedside—but this, she realized, was part of the work.

And though normally in the course of her own life she never went to the beautiful mountains of Rocky Mountain National Park, a two-hour drive southwest of Fort Collins, having no time or impulse for leisure or recreation, she thought that I might like to go, and that this would also allow me to observe her in a quite different context—one in which we could perhaps feel unprogrammed, free.

We piled our stuff into Temple's car—with its four-wheel drive, it was the thing for mountain terrain, especially if we wandered off-road—and took off around nine for the national park. It was a spectacular route: we climbed to higher and higher altitudes on a hairpin road, with terrifying bends, and saw towering cliffs with banded rock strata, foaming gorges far below, and a marvelous range of evergreens, mosses, and ferns. I had the binoculars out constantly and exclaimed at the wonders at every turn.

As we drove on into the park, the landscape opened out into an immense mountain plateau, with limitless views in every direction. We pulled off the road and gazed toward the Rockies—snowcapped, outlined against the horizon, luminously clear even though they were nearly a hundred miles away. I asked Temple if she did not feel a sense of their sublimity. "They're pretty, yes. Sublime, I don't know." When I pressed her, she said that she was puzzled by such words and had spent much time with a dictionary, trying to understand them. She had looked up "sublime," "mysterious," "numinous," and "awe," but they all seemed to be defined in terms of one another.

"The mountains are pretty," she repeated, "but they don't give me a special feeling, the feeling you seem to enjoy." After living for three and a half years in Fort Collins, she said, this was only the second time she had been to them.

What Temple said here seemed to me to have an element of sadness or wistfulness, even of poignancy. She had said similar things on the way up to the park ("You look at the brook, at the flowers, I see what great pleasure you get out of it. I'm denied that"), and, indeed, throughout the weekend. There had been a spectacular sunset the evening before (the sunsets have been particularly fine since Mount Pinatubo erupted), and this, too, she found "pretty" but nothing more. "You get such joy out of the sunset," she said. "I wish I did, too. I know it's beautiful, but I don't 'get' it." Her father, she added, often expressed similar sentiments.

I thought about what Temple had said on Friday night as we walked under the stars. "When I look up at the stars at night, I know I should get a 'numinous' feeling, but I don't. I would like to get it. I can understand it intellectually. I think about the Big Bang, and the origin of the universe, and why we are here: Is it finite, or does it go on forever?"

"But do you get a feeling of its grandeur?" I asked.

"I intellectually understand its grandeur," she replied, and continued, "Who are we? Is death the end? There must be re-ordering forces in the universe. Is it just a Black Hole?"

These were grand words, grand thoughts, and I found myself looking at Temple with a heightened sense of her mental spaciousness, her courage. Or were they, for her, just words, just concepts? Were they purely mental, purely cognitive or intellectual, or did they correspond to any real experience, any passion or feeling?

Now we drove on, higher and higher, the air becoming thinner, the trees smaller, as we moved toward the summit. There was a lake near the park, Grand Lake, which I especially wanted to swim in (I am always excited by the prospect of swimming in exotic, remote lakes: I dream of Lake Baikal and Lake Titicaca), but, sadly, since I had a plane to catch, we did not have time.

On the way back down the mountain, we stopped the car for a brief plant- and bird-spotting geological walk—Temple knew all the plants, all the birds, the geological formations, even though, she said, she had "no special feeling" for them—and then we started the long descent. At one point, just outside the park, seeing a huge, inviting flat sheet of water, I asked Temple to pull over, and impetuously scrambled down toward it: I would have my swim, even though we had not made it to the lake.

It was only when Temple yelled "Stop!" and pointed that I paused in my headlong descent and looked up, and saw that my flat sheet of water, my "lake," so still just in front of me, was accelerating at a terrifying rate a few yards to the left, prior to rushing over a hydroelectric dam a quarter of a mile away. There would have been a fair chance of my being swept along, out of control, right over the dam. There was a look of relief on Temple's face when I stopped and climbed back. Later, she phoned a friend, Rosalie, and said she had saved my life.

We talked of many things on the way back to Fort Collins. Temple mentioned an autistic composer she knew ("He would take bits and pieces of music he had heard, and rearrange them"), and I spoke of Stephen Wiltshire, the autistic artist. We wondered about autistic novelists, poets, scientists, philosophers. Hermelin, who has studied (low-functioning) autistic savants for many years, feels that though they may have enormous talents, they are so lacking in subjectivity and inwardness that major artistic creativity is beyond them. Christopher Gillberg, one of the finest clinical observers of autism, feels that autistic people of the Asperger type, in contrast, may be capable of major creativity and wonders whether indeed Bartók and Wittgenstein may have been autistic. (Many autistic people now like to think of Einstein as one of themselves.)

Temple had spoken earlier of being mischievous, or naughty, saying she enjoyed this at times, and she had been pleased at having smuggled me successfully into the slaughterhouse. She likes to commit small infractions on occasion—"I sometimes walk two feet outside the line at the airport, a little act of defiance"—but all this is in a totally different category from "real badness." That could have terrifying, instantly lethal consequences. "I have a feeling that if I do anything really bad, God will punish me, the steering linkage will go out on the way to the airport," she said as we were driving back. I was startled by the association of divine retribution with a broken steering linkage; I had never thought about how an autistic person, with a wholly causal or scientific view of the universe and a deficient sense of agency or intention, might formulate such matters as divine judgment or will.

Temple is an intensely moral creature. She has a passionate sense of right and wrong, for example, in regard to the treatment of animals; and law, for her, is clearly not just the law of the land but, in some far deeper sense, a divine or cosmic law, whose violation can have disastrous effects—seeming breakdowns in the course of nature itself. "You've read about action at a distance, or quantum theory," she said. "I've always had the feeling that when I go to a meat plant I must be very careful, because God's watching. Quantum theory will get me."

Temple started to become excited. "I want to get this out before you get to the airport," she said, with a sort of urgency.

She had been brought up an Episcopalian, she told me, but had rather early "given up orthodox belief"—belief in any personal deity or intention—in favor of a more "scientific" notion of God. "I believe there is some ultimate ordering force for good in the universe—not a personal thing, not Buddha or Jesus, maybe something like order out of disorder. I like to hope that even if there's no personal afterlife, some energy impression is left in the universe. . . . Most people can pass on genes—I can pass on thoughts or what I write.

"This is what I get very upset at. . . ." Temple, who was driving, suddenly faltered and wept. "I've read that libraries are where immortality lies. . . . I don't want my thoughts to die with me. . . . I want to have done something. . . . I'm not interested in power, or piles of money. I want to leave something behind. I want to make a positive contribution—know that my life has meaning. Right now, I'm talking about things at the very core of my existence."

I was stunned. As I stepped out of the car to say goodbye, I said, "I'm going to hug you. I hope you don't mind." I hugged her—and (I think) she hugged me back.

Selected Bibliography

Choice is always personal and idiosyncratic, and what follows is a selection of sources which I have found enjoyable and intriguing, as well as informative, and which I would encourage the reader to sample. A full reference list follows this section. I have, in addition, listed some favorite or important books to the general reference list, even when no reference has been made to them in the text.

PREFACE

L. S. Vygotsky's early papers, lost for many years, have been recovered and translated into English recently as *The Fundamentals of Defectology*.

In his autobiography, *The Making of Mind*, A. R. Luria traces his own intellectual development in relation to the changing moods of neurology throughout his long lifetime; his chapter on "Romantic Science" particularly brings out his sense of the indispensability of case histories, and how the narrative is crucial to medicine. His own two "romantic" case histories—*The Mind of a Mnemonist* and *The Man with a Shattered World*—are the finest contemporary examples of such histories. A fine critical essay on "inside" narratives of illness is Anne Hunsaker Hawkins's *Reconstructing Illness: Studies in Pathography*.

Kurt Goldstein's general discussion of neurological health, disorder, and rehabilitation is to be found in his remarkable 1939 book, *The Organism* (especially Chapter 10).

The postwar rationalist thinkers on health and disease have been especially Georges Canguilhem and Michel Foucault. Central books are Canguilhem's *The Normal and the Pathological* and Foucault's *Mental Illness and Psychology*.

Gerald Edelman has published five books on his theory of neuronal group selection; the most recent and most readable is *Bright Air, Brilliant Fire*. Israel Rosenfield's *The Invention of Memory* gives a clear history of classical, localizationist neurology, and a sense of how radically neurology may have to be revised in the light of Edelman's theory. I find Edelman's ideas extremely exciting, providing a neural basis, as they aim to do, for the entire range of mental processes from perception to consciousness, and for what it means to be human and a self. An entire new theoretical neuroscience seems to spring from them. I have published two essays on Edelman's work myself in *The New York Review of Books*: "Neurology and the Soul" and "Making Up the Mind."

In a more general way, I have very much enjoyed Freeman Dyson's *Infinite in All Directions* (originally entitled, when given as the Gifford Lectures, "In Praise of Diversity"). The sense of nature's richness and complexity and creativity is also conveyed in all of Ilya Prigogine's books—my favorite is *From Being to Becoming*—and in a book of extraordinary range, Murray Gell-Mann's *The Quark and the Jaguar: Adventures in the Simple and the Complex.*

THE CASE OF THE COLORBLIND PAINTER

A charming early book (it contains the report on the achromatopic surgeon who fell off his horse, and other gems) is Mary Collins's 1925 *Colour-Blindness*. Arthur Zajonc's *Catching the Light: The Entwined History of Light and Mind* is a beautifully researched and written book, especially interesting in its consideration of Goethe's ideas on color and their relation to Land's. (Zajonc also speaks of the case of Jonathan I.)

Though Schopenhauer wrote a youthful essay "On Vision and Colour," this is not readily accessible in English. But thoughts on color vision punctuate his magnum opus, *The World as Will and Representation*, and increased with every edition in his lifetime.

The nineteenth-century debate between different theories of color vision and their advocates comes to life in Steven Turner's *In the Eye's Mind: Vision and the Helmholtz-Hering Controversy*, and in an excellent essay-review of this by C. R. Cavonius.

Semir Zeki has been the pioneer investigator of mechanisms of color perception in the monkey; a synthesis of his work and its relation to current neuroscience is provided in his book *A Vision of the Brain*. A grand synthesis at a higher level, the level of visual awareness, is given by Francis Crick in *The Astonishing Hypothesis: The Scientific Search for the Soul*. Both of these books are quite accessible to the general reader. (And both discuss at length the case of Jonathan I.)

Antonio and Hanna Damasio and their colleagues have published many minute clinical studies of cerebral achromatopsia. Antonio Damasio has given a very full, if somewhat technical, account of this and other visual disorders in his chapter in *Principles of Behavioral Neurology*, and a more general account, coupled with reflections on the theoretical and philosophical importance of such observations, in his recent book, *Descartes' Error*.

Edwin Land's papers have recently been published in their entirety, but one of the most vivid of his accounts is "The Retinex Theory of Color Vision," in *Scientific American*. An excellent essay on Land is "I Am a Camera," by Jeremy Bernstein (this, too, refers to the case of Jonathan I.). And a fascinating film showing the chaos that would result if we did not have color constancy is *Colorful Notions*, originally broadcast by the BBC's Horizon Series in 1984.

The Oxford Companion to the Mind, edited by Richard Gregory, is an indispensable reference on all sorts of neurological and psychological topics. It includes very good articles by Tom Troscianko, "Colour Vision: Brain Mechanisms"; by W. A. H. Rushton, "Colour Vision: Eye Mechanisms"; and by J. J. McCann, "Retinex Theory and Colour Constancy."

An interesting account of the beginnings of color photography, "The First Color Photographs," by Grant B. Romer and Jeannette Delamoir, was published in the *Scientific American* of December 1989. I published a letter on the subject, with reminiscences of color photography in the 1940s, in the March 1990 issue. A centenary article, "Maxwell's Color Photograph," by Ralph M. Evans, appeared in the November 1961 *Scientific American*.

The personal experiences of a congenitally achromatopic man (who is also a vision scientist) are beautifully described in Knut Nordby's "Vision in a Complete Achromat: A Personal Account."

Finally, Frances Futterman, the achromatopic woman whose letters I have excerpted here, has started publishing the *Achromatopsia Network Newsletter* and hopes to network with achromatopic people all over the world. She may be contacted at Box 214, Berkeley, CA 94701-0214.

THE LAST HIPPIE

The grand describer of both frontal lobe and amnesic syndromes was A. R. Luria, in (respectively) *Human Brain and Psychological Processes* and *The Neuropsychology of Memory*. Both of these books are somewhat academic; it was Luria's last wish to supplement them with "romantic" case histories. François Lhermitte's two long papers entitled "Human Autonomy and the Frontal Lobes" give a vivid picture of his sympathetic and naturalistic approach to such patients.

By contrast, the ruthlessness that characterized the lobotomy era is described in a frightening book, *Great and Desperate Cures*, by Elliot Valenstein. A superb essay review of this was written for *The New York Review of Books* by Macdonald Critchley.

The case of Phineas Gage has excited unceasing neurological interest for nearly 150 years and even now is being reexplored using the most sophisticated techniques of reconstructive neuroimaging (see Damasio et al.'s *Science* article). The deepest exploration of the case, and its relevance to all nineteenth-century theorizing about the nervous system from Gall to Freud, has been provided by Malcolm Macmillan in "Phineas Gage: A Case for All Reasons" and by Antonio Damasio in *Descartes' Error*.

Selected Bibliography

Two of my earlier studies on memory, referred to in this chapter—"The Lost Mariner" and "A Matter of Identity"—are reprinted in *The Man Who Mistook His Wife for a Hat.*

The field of memory research is extremely active now, and it is almost invidious to single out names. But Larry Squire and Nelson Butters are certainly leaders in this field and, individually and jointly, have written innumerable papers over the years, as well as edited the volume *The Neuropsychology of Memory.* Other suggested readings on the subject of memory are included in the suggested readings for "The Landscape of His Dreams."

There is also an explosion of interest in the neurology of music and all its therapeutic powers in patients with neurological disorders. Anthony Storr, the psychiatrist, has written a beautiful book, *Music and the Mind,* which touches on every aspect of human response to music. In a chapter entitled "Music and the Brain," in the forthcoming book *Music and Neurologic Rehabilitation,* I have focused more narrowly on the possible ways in which music can affect the brain.

Mickey Hart has written about percussion and rhythm in many cultures, in *Drumming at the Edge of Magic.*

A SURGEON'S LIFE

Gilles de la Tourette's two-part paper, "Étude sur une affection nerveuse," was published in 1885, and a partial translation is included, with a commentary, in "Gilles de la Tourette on Tourette Syndrome," by C. G. Goetz and H. L. Klawans. Meige and Feindel's great book, *Les Tics et leur traitement,* was published in 1902 and translated by Kinnier Wilson in 1907. This book is remarkable not only for its comprehensiveness, but for its tone—the authors' respect for their subjects and the real conversations between them and their physicians. It includes a unique, early autobiographical narrative, "Les Confidences d'un ticqueur."

It is only in the last few years that there have been more accounts from the inside about what it can mean to live with Tourette's. A series of such inside narratives, edited by Adam Seligman and John Hilkevich, was published as *Don't Think About Monkeys.*

I have written a number of papers on Tourette's: "Witty Ticcy Ray," originally published in 1981, was republished in *The Man Who Mistook His Wife for a Hat,* along with "The Possessed." A general overview of the subject is given in "Neuropsychiatry and Tourette's," published in 1989, and more briefly and recently in "Tourette's Syndrome: A Human Condition." A particular aspect of Tourette's that has always fascinated me was presented in "Tourette's and Creativity"; and research on the speed and accuracy of Tourettic movement, "Movement Perturbations Due to Tics," appeared in the 1993 Society for Neuroscience Abstracts.

The Tourette Syndrome Association, 42-40 Bell Boulevard, Bayside, NY

11361, first founded in 1971, disseminates information, gives physician referrals, and sponsors research. It can be contacted at (718) 224-2999 or (800) 237-0717 for information on local chapters.

TO SEE AND NOT SEE

The restoration of vision to those blinded early in life, though rare, has been documented with great care since Cheselden's report in 1728. All known cases up to 1930 are summarized in von Senden's encyclopedic book, *Space and Sight*. Many of these are analyzed by Hebb in his *Organization of Behaviour* and form, along with much other observational and experimental data he provides, crucial evidence that "seeing"—visual perception—must be learned.

The single richest and most detailed case study is that of Richard Gregory and Jean Wallace. This was subsequently reprinted, with further additions, including an exchange of letters with von Senden, in Gregory's *Concepts and Mechanisms of Perception*. The philosophical background to the Molyneux question and the impact of the Cheselden case are also well described by Gregory in his article "Recovery from Blindness," in *The Oxford Companion to the Mind*.

Alberto Valvo's deeply pondered cases of patients submitted to a new surgical procedure for corneal reconstruction are described in his *Sight Restoration after Long-Term Blindness*.

The effects of late blindness—most especially its effects on visual imagery and memory, orientations, and attitudes—have been masterfully described by John Hull in his autobiographical book, *Touching the Rock*. And the restoration of vision after late blindness is finely described in *Second Sight*, by Robert Hine.

One of the deepest, widest-ranging explorations of what it may mean in terms of identity to be blind, both to the individual and to those around him, was given by Diderot in his great *Letter on the Blind: For the Use of Those Who Can See* (he wrote a similar *Letter on the Deaf and Dumb: For the Use of Those Who Can Hear and Speak*). Von Feuerbach's account of Kaspar Hauser contains a remarkable description of his profound visual agnosia when first released into the daylight, after being kept in a lightless dungeon since infancy (pp. 64–5).

These themes have not only been the subject of philosophical discussions and case reports, but of fiction and dramatic reconstruction, ever since Diderot's imagination of Nicholas Saunderson's deathbed. In 1909 the novelist Wilkie Collins based a novel, *Poor Miss Finch*, on such a subject, and the theme is also central in Gide's early novel *La Symphonie pastorale*. A more recent treatment is a brilliant reconstruction by Brian O'Doherty, *The Strange Case of Mademoiselle P.*, very closely based on Mesmer's original 1779 account. In Brian Friel's 1994 play, *Molly Sweeney*, the central character is, like Virgil, blind from early life with retinal damage and cataracts, and, following the removal of the cataracts in middle life, is plunged into a state of agnosic confusion and ambivalence, which is resolved only by a final reversion to blindness.

THE LANDSCAPE OF HIS DREAMS

The original report on Franco Magnani, written by Michael Pearce and illustrated with reproductions of Franco's paintings and Susan Schwartzenberg's photographs in linked pairs, is found in the *Exploratorium Quarterly* for Summer 1988.

Esther Salaman's *A Collection of Moments* provides a beautiful literary and psychological study of "involuntary memories" as they occurred in Proust, Dostoevsky, and other writers. An excerpt from this, and the greater part of Schachtel's paper on memory and childhood amnesia, Stromeyer's classic account of an Eidetiker, a segment of Luria's *Mind of a Mnemonist*, and much else, are to be found in an invaluable sourcebook, Ulrich Neisser's *Memory Observed*.

Frederic Bartlett's classic book, *Remembering*, brings together his experiments showing the constructive, imaginative quality of memory.

The eruption of "experiential" memories during seizures (and their elicitation by direct stimulation of the brain at surgery) is described in almost novelistic detail by Wilder Penfield (and his colleague Perot) in a booklength article, "The Brain's Record of Visual and Auditory Experience," in *Brain*. This same volume of the journal also contains a striking account of Dostoevsky's epilepsy, by Alajouanine. A readable and accessible description of TLE and Dostoevsky syndrome, both in relation to ordinary people and to celebrated artists and thinkers, is given in Eve LaPlante's *Seized: Temporal Lobe Epilepsy as a Medical, Historical, and Artistic Phenomenon.*

A good historical discussion and acute psychoanalytic consideration of nostalgia is given by David Werman in "Normal and Pathological Nostalgia."

PRODIGIES

Darold Treffert's *Extraordinary People* is an excellent introduction to the subject of idiot savants, drawing as it does equally on historical accounts (from Séguin, Down, Tredgold, and others) and Treffert's own clinical experience.

In a more academic vein, *The Exceptional Brain*, edited by Loraine Obler and Deborah Fein, brings together a great range of research regarding human talents in general, and savant talents in particular.

Steven Smith's book, *The Great Mental Calculators*, is the fullest source of observations on calculating talent as it occurs in normal as well as retarded and autistic people.

A particular favorite of mine, never noted by current writers, is F. W. H. Myers's *Human Personality*. Myers himself was a genius, and this shows in every sentence of his great (though often absurd) two-volume book. The chapter on "Genius" is a penetrating and prescient account of computing talents in relation to the cognitive unconscious.

Though Lorna Selfe's *Nadia: A Case of Extraordinary Drawing Ability in an Autistic Child* is, sadly, out of print, Howard Gardner's *Art, Mind, and Brain* contains an important essay on Nadia, which was to some extent the starting point of his subsequent, widely ramifying studies on intelligence and creativity. A particularly thoughtful review of *Nadia* is provided by Clara Claiborne Park, in which she compares Nadia's work with that of her daughter, Jessy, and other autistic artists.

The most detailed cognitive investigation of a musical savant, Eddie, is given by Leon K. Miller in his book *Musical Savants*.

The extensive investigations of Beate Hermelin and her colleagues (including Neil O'Connor and Linda Pring) are mostly available as individual papers, which include detailed studies of Stephen Wiltshire and other savants. An early paper by O'Connor and Hermelin, "Visual and Graphic Abilities of the Idiot Savant Artist," reproduces and discusses some of Stephen's early work.

The 1945 monograph on a savant subject, L., "A Case of 'Idiot Savant': An Experimental Study of Personality Organization," by Martin Scheerer, Eva Rothmann, and Kurt Goldstein, raises fundamental questions unanswered (and often unasked) today. It is, to my mind, the deepest and most searching analysis ever made of the savant (and autistic) mind. L. is clearly autistic, though this term is not used, because the original version of the paper appeared in 1941, before Kanner's description of autism. In their later, fuller 1945 paper, Goldstein et al. compare their formulations with Kanner's.

Merlin Donald's book, *Origins of the Modern Mind*, in which he speculates on the mimetic powers of primitive man, opens vast historical vistas and is one of the most powerfully argued and imaginative reconstructions I have seen of our past (and perhaps future) mental evolution. Jerome Bruner has explored the development of thinking in the child for many years; a very clear account of the "enactive" stage is given in *Studies in Cognitive Growth*.

A fascinating and richly illustrated study of a gifted, retarded octogenarian artist is John MacGregor's *Dwight Macintosh: The Boy Whom Time Forgot*.

I have written three other case histories of savant syndrome, all published in *The Man Who Mistook His Wife for a Hat*: "The Autist Artist," "The Twins," and "A Walking Grove."

Finally, and most importantly, there are Stephen's own books: *Drawings, Cities, Floating Cities*, and *Stephen Wiltshire's American Dream*. (Unfortunately, only *Floating Cities* is currently in print in the United States.)

See the suggested readings for "An Anthropologist on Mars" for more books on autism, and for autism associations.

Selected Bibliography

AN ANTHROPOLOGIST ON MARS

The delineation of autism as a medical condition goes back to the pioneer papers of Kanner, Asperger, and Goldstein in the 1940s; while it was psychiatrically defined (with misleading suggestions of parental etiology) by Bruno Bettelheim in the 1950s (and later in *The Empty Fortress*), and finally established as a biological condition in the 1960s (when Bernard Rimland's *Infantile Autism* was published), autism was not fully portrayed as a human condition until biographical and finally autobiographical narratives began to appear.

One of the first (and still the best) of these is *The Siege: The First Eight Years of an Autistic Child*, by Clara Claiborne Park. Mira Rothenberg's *Children with Emerald Eyes* is a collection of portraits—at once clinical, analytic, empathetic, and poetic—of a dozen children among the hundreds in her pioneering Blueberry Treatment Centers. Charles Hart, in *Without Reason*, provides a remarkable account of his experience of having first an older brother, then a son, with autism. Jane Taylor McDonnell's beautifully written *News from the Border* contains an afterword by her autistic son, Paul.

There has indeed been an explosion of books written about and by autistic people since 1990 (many centering on the complex questions of facilitated communication), and it is difficult to mention any of these without appearing to ignore others. But in terms of its forthrightness, its vigor, its fullness and insight (to say nothing of its priority—for it was the book that gave direct, personal access to an autistic world for the first time), there is nothing to match Temple Grandin's own book, *Emergence: Labeled Autistic*.

Uta Frith's *Autism: Explaining the Enigma* is a very clear and balanced account, though oriented perhaps too exclusively in a "theory of mind" direction. *Autism and Asperger Syndrome*, edited by Frith, contains a number of important articles, including clinical accounts by Christopher Gillberg, Digby Tantam, and Margaret Dewey. It also contains an essay on the autobiographical writings of Asperger adults, including Temple, by Francesca Happé; and the first English translation of Asperger's original 1994 paper, appended to a searching essay by Frith on his contributions. Asperger was, in a sense, "discovered" by Lorna Wing, and her essay comparing his approach and insights with Kanner's also appears in this volume.

The Autism Society of America has chapters throughout the United States and in Puerto Rico. The national headquarters can be contacted at 7910 Woodmont Avenue, Suite 650, Bethesda, MD 20814, telephone (301) 565-0433 or (800) 328-8476. In England, the National Autistic Society is located in 276 Willesden Lane, London NW2 5RB, telephone (081) 451-1114. More Able Autistic People (MAAP), Box 524, Crown Point, IN 46307, publishes a newsletter on higher-functioning people with autism. The Autism Society of Canada is at 129 Yorkville Avenue, Suite 202, Toronto, Ontario M5R 1C4, telephone (416) 922-0302.

References

Alajouanine, T. "Doestoevski's epilepsy." *Brain* 86:209–21 (1963).

Alkon, Daniel L. *Memory's Voice: Deciphering the Brain-Mind Code.* New York: HarperCollins, 1992.

Asperger, Hans. " 'Autistic Psychopathy' in Childhood." In Uta Frith, ed., *Autism and Asperger Syndrome.* New York: Cambridge University Press, 1991.

Bartlett, Frederic C. *Remembering: A Study of Experimental and Social Psychology.* Cambridge: Cambridge University Press, 1932.

Bear, David. "The Neurology of Art: Artistic Creativity in Patients with Temporal Lobe Epilepsy." Paper presented at "The Neurology of Art" symposium, Art Institute of Chicago and Michael Reese Hospital, Chicago, 1988.

―――. "Temporal Lobe Epilepsy: A Syndrome of Sensory-Limbic Hyperconnection." *Cortex* 15:357–84 (1979).

Berkeley, George. *A New Theory of Vision.* 1709. Everyman ed. New York: Dutton, 1910.

Bernstein, Jeremy. "I Am a Camera." In *Cranks, Quarks, and the Cosmos.* New York: Basic Books, 1993.

Bettelheim, Bruno. *The Empty Fortress: Infantile Autism and the Birth of the Self.* New York: Free Press, 1967.

Borges, Jorge Luis. "Funes the Memorious." In *A Personal Anthology.* New York: Grove Press, 1967.

Boyd, Brian. *Vladimir Nabokov: The Russian Years.* Princeton, N.J.: Princeton University Press, 1990, esp. pp. 70–1.

Brann, Eva T. H. *The World of the Imagination: Sum and Substance.* Savage, Md.: Rowman and Littlefield, 1991.

Bruner, Jerome. *Acts of Meaning.* Cambridge: Harvard University Press, 1990.

Bruner, Jerome, and Carol Feldman. "Theories of Mind and the Problem of Autism." In S. Baron-Cohen, H. Tager-Flusberg, and D. J. Cohen, eds. *Understanding Other Minds,* 267–91. New York: Oxford Medical, 1993.

References

Bruner, Jerome S.; Rose R. Olver; Patricia M. Greenfield; et al. *Studies in Cognitive Growth.* New York: Wiley, 1966.

Cahan, David, ed. *Hermann von Helmholtz and the Foundations of Nineteenth-Century Science.* Berkeley: University of California Press, 1993.

Calvin, William H. *The Cerebral Symphony: Seashore Reflections on the Structure of Consciousness.* New York: Bantam Books, 1990.

────── and George A. Ojemann. *Conversations with Neil's Brain: The Neural Nature of Thought and Language.* New York: Addison-Wesley, 1994.

Canguilhem, Georges. *The Normal and the Pathological.* tr. Carolyn Fawcett. New York: Zone Books, 1989.

Cavonius, C. R. "Not Seeing Eye to Eye" (review of *In the Eye's Mind* by R. S. Turner), *Nature* 370:259–60, July 28, 1994.

Chesterton, G. K. *The Secret of Father Brown.* London: Cassell & Co., 1927.

Churchland, Patricia S. *Neurophilosophy: Toward a Unified Science of the Mind-Brain.* Cambridge, Mass.: Bradford Books, MIT Press, 1986.

Coleridge, Samuel Taylor. *Biographia Literaria.* 1817. Reprint, Oxford: Oxford University Press, 1907.

Collins, Mary. *Colour-Blindness.* New York: Harcourt, Brace & Co., 1925.

Collins, Wilkie. *Poor Miss Finch.* New York: Scribner's, 1909.

Crick, Francis. *The Astonishing Hypothesis: The Scientific Search for the Soul.* New York: Scribner's, 1994.

Critchley, E. M. R., ed. *The Neurological Boundaries of Reality.* London: Farrand Press, 1994.

Critchley, Macdonald. "Unkind Cuts." *The New York Review of Books,* April 24, 1986.

Damasio, A.; T. Yamada; H. Damasio; J. Corbett; and J. McKee. "Central Achromatopsia: Behavioral, Anatomic, and Physiologic Aspects." *Neurology* 30, no. 10:1064–71 (October 1980).

Damasio, Antonio R. "Disorders of Complex Visual Processing." In M. Marsel Mesulam, ed., 259–88. *Principles of Behavioral Neurology,* Philadelphia: F. A. Davis Co., 1985.

────── . *Descartes' Error: Emotion, Reason, and the Human Brain.* New York: Grosset/Putnam, 1994.

Damasio, Hanna; Thomas Grabowski; Randall Frank; Albert M. Galaburda; and Antonio Damasio. "The Return of Phineas Gage: Clues about the Brain from the Skull of a Famous Patient." *Science* 264:1102–5, May 20, 1994.

Dennett, Daniel C. *Consciousness Explained.* Boston: Little, Brown and Co., 1991.

Diderot, Denis. *Lettre sur les aveugles.* Paris: Durand, 1749.

────── . *Lettre sur les sourds et muets.* Paris, 1751.

Donald, Merlin. *Origins of the Modern Mind: Three Stages in the Evolution of Culture and Cognition.* Cambridge: Harvard University Press, 1991.

Donaldson, Margaret. *Human Minds: An Exploration.* London: Allen Lane, 1992.

Down, J. Langdon. *Mental Affections of Childhood and Youth*. 1887. Facsimile ed., Oxford: MacKeith Press, Blackwell Scientific Publications, 1990.

Dyson, Freeman J. *Infinite in All Directions*. New York: Harper & Row, 1988.

Edelman, Gerald M. *Bright Air, Brilliant Fire: On the Matter of the Mind*. New York: Basic Books, 1992.

————. *Neural Darwinism: The Theory of Neuronal Group Selection*. New York: Basic Books, 1987.

————. *The Remembered Present: A Biological Theory of Consciousness*. New York: Basic Books, 1989.

————. *Topobiology: An Introduction to Molecular Embryology*. New York: Basic Books, 1988.

Edelman, Gerald M., and Vernon B. Mountcastle. *The Mindful Brain*. Cambridge: MIT Press, 1978.

Edridge-Green, F. W. *The Physiology of Vision: with Special Reference to Colour Blindness*. London: G. Bell and Sons, 1920.

Evans, Ralph M. "Maxwell's Color Photograph." *Scientific American*, November 1961.

Feldman, David H., with Lynn T. Goldsmith. *Nature's Gambit: Child Prodigies and the Development of Human Potential*. New York: Basic Books, 1986.

Feuerbach, Anselm von. *Caspar Hauser: An Account of an Individual Kept in a Dungeon, Separated from all Communication with the World, from Early Childhood to about the Age of Seventeen*. 1832. English trans., London: Simpkin & Marshall, 1834.

Foucault, Michel. *Mental Illness and Psychology*. Berkeley: University of California Press, 1987.

Freud, Sigmund. "Constructions in Analysis." *International Journal of Psychoanalysis* 19:377 (1938). Reprinted in Philip Rieff, ed., *Therapy and Technique*, New York: Collier, 1963.

Friel, Brian. *Molly Sweeney*. Old Castle, Co. Meath, Ireland: The Gallery Press, 1994.

Frith, Uta. *Autism: Explaining the Enigma*. New York: Blackwell, 1989.

Frith, Uta, ed. *Autism and Asperger Syndrome*. New York: Cambridge University Press, 1991.

Fuller, G. N., and M. V. Gale, "Migraine Aura as Artistic Inspiration." *British Medical Journal* 297:1670–72 (December 24, 1988).

Gardner, Howard. *Art, Mind, and Brain: A Cognitive Approach to Creativity*. New York: Basic Books, 1982.

————. *Frames of Mind: The Theory of Multiple Intelligences*. New York: Basic Books, 1983.

Gastaut, Henri. "Memories originaux: la maladie de Vincent van Gogh envisagée à la lumière des conceptions nouvelles sur l'epilepsie psychomotrice." *Annales Medico-Psychologiques* 114:196–238, 1956.

Gazzaniga, Michael S. *Nature's Mind: The Biological Roots of Thinking, Emotions, Sexuality, Language, and Intelligence*. New York: Basic Books, 1992.

References

Gell-Mann, Murray. *The Quark and the Jaguar: Adventures in the Simple and the Complex*. New York: Freeman, 1994.

Geschwind, Norman. "Epilepsy in the Life and Writings of Dostoievsky." Lecture, Boston Society of Psychiatry and Neurology, March 16, 1961.

Gide, André. *La Symphonie pastorale*. 1919. Reprint, London: Penguin Books, 1963.

Gillberg, Christopher. "Clinical and Neurobiological Aspects of Asperger Syndrome in Six Family Studies." In Uta Frith, ed., *Autism and Asperger Syndrome*. New York: Cambridge University Press, 1991.

Gillberg, Christopher, and Mary Coleman. *The Biology of the Autistic Syndromes*. 2nd ed. New York: MacKeith Press, Cambridge University Press, 1992.

Goethe, J. W. von. "Theory of Color." In Douglas Miller, ed. and trans., *Scientific Studies*, 164. Vol. 12, *Goethe: Collected Works in English*. New York: Suhrkamp, 1988.

Goldberg, Elkhonon, and William B. Barr. "Three Possible Mechanisms of Unawareness of Deficit." In G. P. Prigatano and D. L. Schachter, eds., *Awareness of Deficit after Brain Injury: Clinical and Theoretical Issues*. New York: Oxford University Press, 1991.

Goldstein, Kurt. *Language and Language Disturbances: Aphasic Symptom Complexes and their Significance for Medicine and Theory of Language*. New York: Grune & Stratton, 1948.

————. *The Organism: A Holistic Approach to Biology Derived from Pathological Data in Man*. 1939. Reprint, New York: Zone Books, MIT Press, 1995.

Gowers, W. R. *Subjective Sensations of Sight and Sound: Abiotrophy, and Other Lectures*. Philadelphia: P. Blakiston's Son & Co., 1904.

Grandin, Temple. "Behavior of Slaughter Plant and Auction Employees Toward the Animals." *Anthrozoos* 1, no. 4: 205–13 (Spring 1988).

————. "An Inside View of Autism." In Eric Schopler and Gary B. Mesibov, eds. *High-Functioning Individuals with Autism*, 105–26. New York: Plenum Press, 1992.

————. "My Experiences as an Autistic Child and Review of Selected Literature." *Journal of Orthomolecular Psychiatry* 13, no. 3: 144–74 (3rd quarter 1984).

————. "Needs of High Functioning Teenagers and Adults with Autism." *Focus on Autistic Behavior* 5, no. 1: 1–16 (April 1990).

Grandin, Temple, and Margaret Scariano. *Emergence: Labeled Autistic*. Novato, Calif.: Arena Press, 1986.

Gregory, R. L., and Jean G. Wallace. "Recovery from Early Blindness: A Case Study." *Quarterly Journal of Psychology* (1963). Reprinted in *Concepts and Mechanisms of Perception*, by R. L. Gregory. London: Duckworth, 1974.

Gregory, Richard L. "Blindness, Recovery from." In Richard L. Gregory, ed., *The Oxford Companion to the Mind*, 94–96. Oxford: Oxford University Press, 1987.

Happé, Francesca G. E. *Autism: An Introduction to Psychological Theory.* London: UCL Press, 1994.

————. "The Autobiographical Writings of Three Asperger Syndrome Adults." In Uta Frith, ed., *Autism and Asperger Syndrome.* New York: Cambridge University Press, 1991.

Hart, Charles. *Without Reason: A Family Copes with Two Generations of Autism.* New York: Harper & Row, 1989.

Hart, Mickey. *Drumming at the Edge of Magic: A Journey into the Spirit of Percussion.* San Francisco, HarperCollins, 1990.

Hawkins, Anne Hunsaker. *Reconstructing Illness: Studies in Pathography.* W. Lafayette, Ind.: Purdue University Press, 1993.

Hebb, D. O. *The Organization of Behaviour.* New York: Wiley, 1949.

Helmholtz, Hermann von. *On Thought in Medicine (Das Denken in der Medizin),* 1878. Reprint, Baltimore: John Hopkins Press, 1938.

————. *Physiological Optics.* 1909. Ed. J. P. C. Southall. Optical Society of America, 1924.

Hermelin, Beate, and Neil O'Connor. "Art and Accuracy: The Drawing Ability of Idiot-Savants." *Journal of Child Psychology and Psychiatry* 31, no. 2, 217–28 (1990).

Hine, Robert V. *Second Sight.* Berkeley: University of California Press, 1993.

Hughlings Jackson, John. "On a Particular Variety of Epilepsy ('Intellectual Aura')." *Brain* 3:179–207 (1880).

Hull, John M. *Touching the Rock: An Experience of Blindness.* New York: Pantheon Books, 1990.

Hurlburt, R. T.; F. Happé; and U. Frith. "Sampling the Form of Inner Experience in Three Adults with Asperger Syndrome." *Psychological Medicine* 24:385–95 (1994).

Jamison, Kay Redfield. *Touched with Fire: Manic-Depressive Illness and the Artistic Temperament.* New York: Free Press, 1993.

Kanner, L. "Autistic Disturbances of Affective Contact." *Nervous Child* 2:217–50 (1943).

Kierkegaard, Søren. *Stages on Life's Way.* 1843. Trans. Walter Lowrie. Princeton, N.J.: Princeton University Press, 1940.

Kosslyn, Stephen M., and Olivier Koenig. *Wet Mind: The New Cognitive Neuroscience.* New York: The Free Press, 1992.

Kremer, Richard L. "Innovation through Synthesis: Helmholtz and Color Research." In David Cahan, ed., *Hermann von Helmoltz and the Foundations of Nineteenth-Century Science.* Berkeley: University of California Press, 1993.

Land, Edwin H. "The Retinex Theory of Color Vision." *Scientific American,* December 1977, 108–28.

Lane, Harlan. *The Mask of Benevolence: Disabling the Deaf Community.* New York: Alfred A. Knopf, 1992.

LaPlant, Eve. *Seized: Temporal Lobe Epilepsy as a Medical, Historical, and Artistic Phenomenon.* New York: HarperCollins, 1993.

Lashley, Karl. "In Search of the Engram." *Symp. Soc. Exp. Biol.* 4:454–82 (1950).

References

Lhermitte, F. "Human Autonomy and the Frontal Lobes." *Annals of Neurology* 19, no. 4:326–43 (April 1986).

Llinás, R. R., and D. Paré. "On Dreaming and Wakefulness." *Neuroscience* 44, no. 3:521–35, 1991.

Locke, John. *Essay Concerning Human Understanding.* 1690. Ed. P. H. Nidditch. Oxford: Oxford University Press, 1975.

Lucci, Dorothy; Deborah Fein; Adele Holevas; and Edith Kaplan. "Paul: A Musically Gifted Autistic Boy." In Loraine Obler and Deborah Fein, eds., *The Exceptional Brain*, 310–24. New York: Guilford Press, 1988.

Luria, A. R. *Human Brain and Psychological Processes.* New York: Harper & Row, 1966.

————. *The Making of Mind: A Personal Account of Soviet Psychology.* Ed. Michael Cole and Sheila Cole. Cambridge: Harvard University Press, 1979.

————. *The Man with a Shattered World.* 1972. Reprint, Cambridge: Harvard University Press, 1987.

————. *The Mind of a Mnemonist.* 1968. Reprint, Cambridge: Harvard University Press, 1987.

————. *The Neuropsychology of Memory.* New York: John Wiley & Sons, 1976.

McCann, J. J. "Retinex Theory and Colour Constancy." In Richard L. Gregory, ed., *The Oxford Companion to the Mind*, 684–85. Oxford: Oxford University Press, 1987.

McDonnell, Jane Taylor. *News from the Border: A Mother's Memoir of Her Autistic Son.* Afterword by Paul McDonnell. New York: Ticknor & Fields, 1993.

MacGregor, John. *Dwight Macintosh: The Boy Whom Time Forgot.* Oakland, Calif.: Creative Growth Art Center, 1992.

McKendrick, John Gray. *Hermann Ludwig Ferdinand von Helmholtz.* London: T. Fisher Unwin, 1899.

McKenzie, Ivy. "Discussion on Epidemic Encephalitis." *British Medical Journal*, September 24, 1927, pp. 632-634.

Macmillan, Malcolm. "Inhibition and the Control of Behavior: From Gall to Freud via Phineas Gage and the Frontal Lobes." *Brain and Cognition* 19:72–104 (1992).

————. "Phineas Gage: A Case for All Reasons." In C. Code, C.W. Wallesch, A. R. Lecours, and Y. Joanette, eds., *Classic Cases in Neuropsychology.* London: Erlbaum, 1995.

Meige, H. and E. Feindel. *Les Tics et leur traitment.* Paris: Masson, 1902. Trans. S. A. Kinnier Wilson as *Tics and Their Treatment.* New York: William Wood & Co., 1907.

Mesulam, M.-Marsel. *Principles of Behavioral Neurology.* Philadelphia: F. A. Davis Co., 1985.

Miller, Leon K. *Musical Savants: Exceptional Skill in the Mentally Retarded.* Hilldale, N.J.: Erlbaum, 1989.

Modell, Arnold H. *Other Times, Other Realities: Toward a Theory of Psychoanalytic Treatment*. Cambridge: Harvard University Press, 1990.

————. *The Private Self*. Cambridge: Harvard University Press, 1993.

Mollon, J. D.; F. Newcombe; P. G. Polden; and G. Ratcliff. "On the Presence of Three Cone Mechanisms in a Case of Total Achromatopsia." In G. Verriest, ed., *Colour Vision Deficiencies*, 130–35. Bristol: Hilger, 1980.

Moreau, J.-J. *Hashish and Mental Illness*. 1845. Reprint, New York: Raven Press, 1973.

Murray, T. J. "Illness and Healing: The Art of Robert Pope." *Humane Medicine* 10, no. 3, July 1994, 199–208.

Myers, Frederic W. H. *Human Personality and Its Survival of Bodily Death*. 2 vols. New York: Longmans, Green & Co., 1903.

Neisser, Ulrich. *Memory Observed: Remembering in Natural Contexts*. San Francisco: Freeman, 1982.

Neisser, Ulrich, and Eugene Winograd, eds. *Remembering Reconsidered: Ecological and Traditional Approaches to the Study of Memory*. Cambridge: Cambridge University Press, 1988.

Nordby, Knut. "Vision in a Complete Achromat: A Personal Account." In R. F. Hess, L. T. Sharpe, and K. Nordby, eds., *Night Vision: Basic, Clinical and Applied Aspects*. Cambridge: Cambridge University Press, 1990.

Nuland, Sherwin. *Doctors: The Biography of Medicine*. New York: Alfred A. Knopf, 1988.

Obler, Loraine K., and Deborah Fein, eds. *The Exceptional Brain: Neuropsychology of Talent and Special Abilities*. New York: Guilford Press, 1988.

O'Connor, N., and B. Hermelin. "Visual and Graphic Abilities of the Idiot Savant Artist." *Psychological Medicine* 17:79–90 (1987).

O'Doherty, Brian. *The Strange Case of Mademoiselle P.* New York: Pantheon Books, 1992.

Park, Clara Claiborne. *The Siege: The First Eight Years of an Autistic Child*. Rev. ed., Boston: Little, Brown, 1982.

————. Review of *Nadia* by Lorna Selfe. *Journal of Autism and Childhood Schizophrenia* 8:457–72 (1978).

Pavlov, Ivan P. *Lectures on Conditioned Reflexes: Twenty-five Years of Objective Study of the Higher Nervous Activity (Behaviour) of Animals*. Trans. W. Horsley Gantt. New York: International Publishers, 1928.

Pearce, Michael. "A Memory Artist." *Exploratorium Quarterly* 12, issue 2, "Memory": 12–17 (Summer 1988).

Penfield, W., and P. Perot. "The Brain's Record of Visual and Auditory Experience: A Final Summary and Discussion." *Brain* 86:595–696 (1963).

Poppel, Ernst. *Mindworks: Time and Conscious Experience*. New York: Harcourt Brace Jovanovich, 1988.

Posner, Michael I., and Marcus E. Raichle. *Images of Mind*. New York: Scientific American Library, 1994.

References

Prigogine, Ilya. *From Being to Becoming*. San Francisco: Freeman, 1980.

Prigogine, Ilya, and Isabelle Stengers. *Order Out of Chaos: Man's New Dialogue with Nature*. New York: Bantam, 1984.

Pring, Linda, and Beate Hermelin. "Bottle, Tulip and Wineglass: Semantic and Structural Picture Processing by Savant Artists," *Journal of Child Psychology and Psychiatry*, in press.

Proust, Marcel. *Remembrance of Things Past*, vol. 11, *The Sweet Cheat Gone*. London: Chatto & Windus, 1949, esp. p. 185.

Ramachandran, V. S. "Behavioral and Magnetoencephalographic Correlates of Plasticity in the Adult Human Brain," *Proceedings of the National Academy of Science* 90:10413–20 (1993).

Rimland, Bernard. *Infantile Autism: The Syndrome and Its Implications for a Neural Theory of Behavior*. New York: Appleton-Century-Crofts, 1964.

Rimland, Bernard, and Deborah Fein. "Special Talents of Autistic Savants." In Loraine Obler and Deborah Fein, eds., *The Exceptional Brain*, 474–92. New York: Guilford, 1988.

Ritvo, Edward; Anne M. Brothers; B.J. Freeman; and Carmen Pingree. "Eleven Possibly Autistic Parents," *Journal of Autism and Developmental Disorders* 18, no. 1, 139 (1988).

Ritvo, Edward R.; Riva Ritvo; B. J. Freeman; and Anne Mason-Brothers. "Clinical Characteristics of Mild Autism in Adults," *Comprehensive Psychiatry* 35, no. 2, 149–56 (March/April 1994).

Rizzo, M.; M. Nawrot,; R. Blake; and A. Damasio. "A Human Visual Disorder Resembling Area V4 Dysfunction in the Monkey." *Neurology* 42:1175–80 (June 1992).

Romer, Grant B., and Jeannette Delamoir. "The First Color Photographs." *Scientific American*, December 1989, 88–96.

Rose, Steven. *The Conscious Brain*. New York: Alfred A. Knopf, 1973; revised ed. 1989.

Rosenfield, Israel. *The Invention of Memory: A New View of the Brain*. New York: Basic Books, 1988.

_____.*The Strange, Familiar, and Forgotten: An Anatomy of Consciousness*. New York: Alfred A. Knopf, 1992.

Rothenberg, Mira. *Children with Emerald Eyes: Histories of Extraordinary Boys and Girls*. New York: Dial Press, 1977.

Rushton, W. A. H. "Colour Vision: Eye Mechanisms." In Richard L. Gregory, ed., *The Oxford Companion to the Mind*, 152–54. Oxford: Oxford University Press, 1987.

Sacks, Oliver. *A Leg to Stand On*. 1984. Rev. ed., New York: HarperCollins, 1993.

_____. *Awakenings*. 1973. Rev. ed., New York: HarperCollins, 1990.

_____. "Color Photography in the Forties." Letter to the editor. *Scientific American*, March 1990.

_____. "The Lost Mariner," in *The Man Who Mistook His Wife for a Hat*, 23–42. New York: HarperCollins, 1985.

———. "Making Up the Mind." *New York Review of Books*, April 8, 1993.

———. *The Man Who Mistook His Wife for a Hat*. New York: HarperCollins, 1985.

———. "A Matter of Identity," in *The Man Who Mistook His Wife for a Hat*, 108–19. New York: HarperCollins, 1985.

———. *Migraine*. 1970. Rev. ed., Berkeley: University of California Press, 1993.

———. "Music and the Brain." In Concetta Tomaino, ed., *Music and Neurologic Rehabilitation*. St. Louis: MMB Music, in press.

———. "Neurology and the Soul." *New York Review of Books*, November 22, 1990.

———. "Neuropsychiatry and Tourette's." In J. Mueller, ed., *Neurology and Psychiatry: A Meeting of Minds*. Basel: S. Karger, 1989.

———. "Tourette's and Creativity." *British Medical Journal* 305:1515–16 (December 19, 1992).

———. "Tourette's Syndrome: A Human Condition." In Roger Kurlan, ed., *Handbook of Tourette's Syndrome and Related Tic and Behavioral Disorders*, 509–14. New York: Marcel Dekker, 1993.

———. "Witty Ticcy Ray," in *The Man Who Mistook His Wife for a Hat*, 92–101. New York: HarperCollins, 1985.

Sacks, Oliver, and Robert Wasserman. "The Case of the Colorblind Painter." *New York Review of Books*, November 19, 1987.

Sacks, O.; O. Fookson; M. Berkenblit; B. Smetanin; and R. M. Siegel. "Movement Perturbations due to Tics." *Society for Neuroscience Abstracts* 19:549 (1993).

Sacks, O.; R. L. Wasserman; S. Zeki; and R. M. Siegel. "Sudden Colorblindness of Cerebral Origin." *Society for Neuroscience Abstracts* 14:1251 (1988).

Salaman, Esther. *A Collection of Moments: A Study of Involuntary Memories*. London: Longman, 1970.

———. *The Great Confession: From Aksakov and De Quincey to Tolstoy and Proust*. London: Allen Lane, 1973.

Schachtel, Ernest G. "On Memory and Childhood Amnesia." *Psychiatry* 10:1–26 (1947).

Scheerer, Martin; Eva Rothmann; and Kurt Goldstein. "A Case of 'Idiot Savant': An Experimental Study of Personality Organization." In John F. Dashiell, ed., *Psychological Monographs* 58, no. 4; 1–63. Evanston: American Psychological Association, 1945.

Schopenhauer, Arthur. *The World as Will and Representation* (1818–1819). 2 vols. Trans. E. F. J. Payne. New York: Dover, 1969.

Schopler, Eric, and Gary B. Mesibov, eds. *High-Functioning Individuals with Autism*. New York: Plenum Press, 1992.

Séguin, Edouard. *Idiocy and Its Treatment by the Physiological Method*. 1866. Reprint, New York: Kelley, 1971.

Selfe, Lorna. *Nadia: A Case of Extraordinary Drawing Ability in an Autistic Child*. London: Academic Press, 1977.

References

Seligman, Adam, and John Hilkevich, eds. *Don't Think About Monkeys*. Duarte, Calif.: Hope Press, 1992.

Sharpe, Lindsay T., and Knut Nordby. "Total Colorblindness: An Introduction." In R. F. Hess, L. T. Sharpe, and K. Nordby, eds., *Night Vision: Basic, Clinical and Applied Aspects*. Cambridge: Cambridge University Press, 1990.

Smith, Steven B. "Calculating Prodigies." In Loraine Obler and Deborah Fein, eds., *The Exceptional Brain*, 19–47. New York: Guilford Press, 1988.

―――. *The Great Mental Calculators: The Psychology, Methods, and Lives of Calculating Prodigies, Past and Present*. New York: Columbia University Press, 1983.

Squire, L. R. *Memory and Brain*. New York: Oxford University Press, 1987.

Squire, L. R., and N. Butters, eds. *The Neuropsychology of Memory*. New York: Guilford Press, 1984.

Stern, Daniel N. *The Interpersonal World of the Infant: A View from Psychoanalysis and Development Psychology*. New York: Basic Books, 1985.

Storr, Anthony. *Music and the Mind*. New York: Free Press, 1992.

Strachey, Lytton. "The Life, Illness, and Death of Dr. North." In *Portraits in Miniature*, 29–39. London: Chatto & Windus, 1931.

Thelen, Esther, and Linda B. Smith. *A Dynamics System Approach to the Development of Cognition and Action*. Cambridge, Mass.: MIT Press, 1994.

Tourette, Georges Gilles de la. "Étude sur une affection nerveuse caractérisée par de l'incoordination motrice accompagnée d'écholalie et de copralalie." *Arch. Neur.* 9 Paris, 1885. A partial translation is included in C. G. Goetz and H. L. Klawans, "Gilles de la Tourette on Tourette syndrome." In A. J. Friedhoff and T. N. Chase, eds., *Advances in Neurology*. Vol. 35, *Gilles de la Tourette Syndrome*. New York: Raven Press, 1982.

Tredgold, A. F. *A Text-Book of Mental Deficiency*. 1908. Reprint, London: Bailliere, Tindall & Cox, 1952.

Treffert, Darold A. *Extraordinary People: An Exploration of the Savant Syndrome*. New York: Harper and Row, 1989; London: Bantam, 1989.

Troscianko, Tom. "Colour Vision: Brain Mechanisms." In Richard L. Gregory, ed., *The Oxford Companion to the Mind*, 150–52. Oxford: Oxford University Press, 1987.

Turner, R. Steven. "Consensus and Controversy: Helmholtz on the Visual Perception of Space." In David Cahan, ed., *Hermann von Helmholtz and the Foundations of Nineteenth-Century Science*. Berkeley: University of California Press, 1993.

―――. *In the Eye's Mind: Vision and the Helmholtz-Hering Controversy*. Princeton, N.J.: Princeton University Press, 1994.

Valenstein, Elliot S. *Great and Desperate Cures: The Rise and Decline of Psychosurgery and Other Radical Treatments for Mental Illness*. New York: Basic Books, 1986.

Valvo, Alberto. *Sight Restoration after Long-Term Blindness: The Problems and Behavior Patterns of Visual Rehabilitation.* New York: American Foundation for the Blind, 1971.

van den Bergh, Sidney, Robert D. McClure, and Robert Evans. "The Supernova Rate in Shapley-Ames Galaxies," *The Astrophysical Journal* 323:44–53 (December 1, 1987).

von Senden, M. *Space and Sight: The Perception of Space and Shape in the Congenitally Blind Before and After Operation.* 1932. Reprint, Glencoe, Ill.: Free Press, 1960.

Vygotsky, L. S. *The Fundamentals of Defectology.* Trans. Jane E. Knox and Carol B. Stevens. In Robert W. Rieber and Aaron S. Carton, eds., *The Collected Works of L. S. Vygotsky.* New York: Plenum, 1993.

Wai-Ching Ho, ed. *Yani: The Brush of Innocence.* New York: Hudson Hills Press, 1989.

Waterhouse, Lynn. "Extraordinary Visual Memory and Pattern Perception in an Autistic Boy." In Loraine Obler and Deborah Fein, eds., *The Exceptional Brain,* 325–38. New York: Guilford Press, 1988.

Waxman, Stephen G., and Norman Geschwind. "Hypergraphia in Temporal Lobe Epilepsy." *Neurology* 24:629–36 (1974).

————. "The Interictal Behavior Syndrome Associated with Temporal Lobe Epilepsy." *Archives of General Psychiatry* 32:1580–86 (1975).

Wells, H. G. "The Country of the Blind." London: Nelson, 1910.

Werman, David S. "Normal and Pathological Nostalgia." *Journal of the American Psychoanalytic Association* 25:387–95 (1977).

Williams, Donna. *Nobody Nowhere.* New York: Times Books, 1992.

————. *Somebody Somewhere.* New York: Times Books, 1994.

Wiltshire, Stephen. *Cities.* London: J. M. Dent, 1989.

————. *Drawings.* London: J. M. Dent, 1987.

————. *Floating Cities.* London: Michael Joseph, 1991; New York: Summit, 1991.

————. *Stephen Wiltshire's American Dream.* London: Michael Joseph, 1993.

Wing, Lorna. "The Relationship between Asperger's Syndrome and Kanner's Autism." In Uta Frith, ed., *Autism and Asperger Syndrome.* New York: Cambridge University Press, 1991.

Yates, Frances. *The Art of Memory.* London: Penguin, 1969.

Young, Thomas. "The Bakerian Lecture: On the Theory of Lights and Colours." *Philosophical Transactions of the Royal Society,* London 92:12–48.

Zajonc, Arthur. *Catching the Light: The Entwined History of Light and Mind.* New York: Bantam, 1993.

Zeki, Semir. *A Vision of the Brain.* Oxford: Blackwell Scientific Publications, 1993.

Zihl, J.; D. Von Cramon; and N. Mai. "Selective Disturbance of Movement Vision after Bilateral Brain Damage." *Brain* 106:313–40, 1983.

Zuckerkandl, Victor. *Sound and Symbol.* 2 vols. Princeton, N.J.: Princeton University Press, 1973.

Index

Index

Casson, Sir Hugh, 203, 206
cat, Virgil's perception of, 121–3
cataracts, 108, 111, 142; surgery for,
    110 and n., 113, 115, 135
cattle behavior, 265–6, 267–8,
    279–80, 281
cerebral palsy, 249n.
Cézanne, Paul, 128
changelings, 56, 190
Charcot, Jean Martin, 77
Chase, Liz, 237, 238
Cheselden, William, 110, 122, 130,
    151
Chesterton, G. K., xix, 183
child prodigies, 191; case history of
    Stephen Wiltshire, 196–222,
    224, 228–43; see also savants
children: cognitive growth in, xvii,
    241n.; perceptual
    development of, 140n., 140–1
chorea, 102
chromatophenes, 28 and n.
Cities (drawings of Stephen
    Wiltshire), 208
Claparède, Edouard, 53n.
Cocteau, Jean, 73n.
coenesthesia, 162, 254n.
cognitive psychology, 245, 246
Cole, Lorraine, 197–8, 202, 210, 237
Coleridge, Samuel, 227, 288n.
Collins, Mary, 8n.
color: agnosia, 17, 126; anomia, 17,
    36n., 126; constancy, 22–3;
    construction, 25–9; cultural
    categories, 17n.; expectation
    and, 29n.; and identity, 35;
    mixing, 23; perception, 125–6,
    147 and n.; photography, 23
    and n., 24; primary, 23;
    theory, 4, 21, 22, 24, 25, 26;
    vision and drugs, 31n.
colorblindness, cerebral (total), 7n.;
    advantages of, 38 and n.; and
    anoxia, 31 and n.; and art,
    13–15, 33, 39–40; awareness
    of, 13n.; beauty of, 38–9;

brain damage and, 19, 30–1,
    31n.; camouflage and, 38n.;
    case history of, 3–41; cerebral
    changes secondary to, 40–1;
    congenital, 9n., 21n., 28n.,
    33n., 34n., 38n.; and contrast
    perception, 9, 32n., 33, 38 and
    n.; and dreams, 11, 12;
    historical cases of, 8n., 19–29;
    horror of, 7, 10–11, 35;
    hysterical or pretended, 16,
    17; identity and, 35–40;
    indescribability of, 7, 10–11,
    35; motion vision and, 38,
    38n.; night vision and, 37n.,
    37–8; sense of loss in, 12–13,
    33; and synesthesia, 11; tests
    for, 15–17, 16n.; transient,
    31n.; visual memory and
    imagery in, 7, 13, 13n., 17,
    35–36, 36n., 40; visual
    migraines and, 11
colorblindness, retinal (red-green), 4,
    9n., 18n., 20n., 24n., 28n.,
    29n., 33n., 34n., 38n.
Colorful Notions (film), 299
compulsions: in autism, 191, 198,
    247, 250, 254, 277, 292; in
    postencephalitic syndromes,
    86n.; in Tourette's syndrome,
    82, 86, 101 and n.
concrete thought, 129n., 155, 218,
    283
consciousness, doubling of, 165
construction: of color, 25–9; of
    motion, 26, 27n.
contrast perception, 9, 32n., 33
coprolalia, 77
coypu, 232–4
creativity, 176n., 241–2
Crick, Francis, 31, 34n.
Critchley, Macdonald, 63

Dalton, John, 18n.
Damasio, Antonio, 7n., 27n., 34,
    282n., 288n.

Index

Index

PERMISSIONS
ACKNOWLEDGMENTS

Grateful acknowledgment is made to the following for permission to reprint previously published material:

Cambridge University Press and *Knut Nordby*: Excerpts from article by Knut Nordby from *Night Vision: Basic, Clinical and Applied Aspects*, edited by R. F. Hess, L. T. Sharpe, and K. Nordby, copyright © 1990 by Cambridge University Press. Reprinted by permission of Cambridge University Press and Knut Nordby.

Farrar, Straus & Giroux, Inc.: Excerpt from "Memories of West Street and Lepke" from *Life Studies* by Robert Lowell, copyright © 1958, 1959 by Robert Lowell, copyright renewed 1981, 1986, 1987 by Harriet W. Lowell, Caroline Lowell, and Sheridan Lowell. Reprinted by permission of Farrar, Straus & Giroux, Inc.

Richard Gregory: Excerpt from sight restoration case history by Richard Gregory with Jean G. Wallace (*The Quarterly Journal of Psychology*, 1963, and reprinted in *Concepts and Mechanisms of Perception* by Richard Gregory). Reprinted by permission of the author.

Grove/Atlantic Publishing: Excerpt from "Funes the Memorious" from *A Personal Anthology* by Jorge Luis Borges (Grove Press, 1967). Reprinted by permission of Grove/Atlantic Publishing.

Ice Nine Publishing Co., Inc.: Excerpt from "Box of Rain," words by Robert Hunter, music by Phil Lesh, performed by Grateful Dead, copyright © 1980 by Ice Nine Publishing Co., Inc. (ASCAP). Reprinted by permission of Ice Nine Publishing Co., Inc.

Oxford University Press: Excerpt from "Selective Disturbance of Movement Vision After Bilateral Brain Damage" by J. Zihl, D. Von Cramon, and N. Mai (*Brain*, 106: 313–340, 1983). Reprinted by permission of Oxford University Press, Oxford, England.

PRO-ED Journals: Excerpts from "Needs of High Functioning Teenagers and Adults with Autism (Tips from a Recovering Autistic)" by Temple Grandin (*Focus on Autistic Behavior*, vol. 5, no. 1, April 1990, pp. 1–16), copyright © 1990 by PRO-ED, Inc. Reprinted by permission of PRO-ED Journals.

PHOTOGRAPHIC CREDITS

Lowell Handler: Mr. I.'s grey fruit, Mr. I.'s grey boat
Daniel G. Hill: Mr. I.'s sunset postcard, Stephen Wiltshire's Matisse heads,
   second drawing of house
Susan Schwartzenberg: Franco Magnani's paintings, Pontito.
Mark Sheinkman: Mr. I.'s paintings

Grateful acknowledgment is made to the artists for permission to reproduce works by Jonathan I. and Franco Magnani; and to J. M. Dent & Sons
Ltd., John Johnson Ltd., and Michael Joseph for permission to reproduce
works by Stephen Wiltshire.